I0481463

How to Build
an Advanced Flood Control System

A Step by Step Guide for Building the Most Effective
Flood Control System for Your Region

Mark Fennell

© 2018

Preface

This book is your Guide to building the most sophisticated Flood Control System possible for your region. In this book you will find step by step instructions for designing and constructing a highly Advanced Flood Control System.

This Advanced Flood Control System will eliminate all flooding within a 150 mile diameter region. It will provide ample drinking water for everyone. It will also provide emergency electrical power, when storms knock out the existing power infrastructure.

The concepts behind this design have been explained in previous books. We assume you understand those basic concepts, and now you are ready to build the system in your area.

The book in your hands is therefore a detailed set of instructions, a step by step guide, for actually constructing the system. Using this Guide, you will have everything you need to construct the Advanced Flood Control System effectively, and efficiently.

75 Full-Color Drawings

The Designer has created numerous full-color drawings. (The current total is 75 pictures). These drawings show all aspects of the system.

This includes: layout; tunnels; piping; containers; digging operations; inter-connection pieces; filters; pumps; hydropower systems; and electrical wiring. This also includes many details on the actual construction process.

However, in order make this Guide shorter (and less expensive) there are no actual pictures in this guide book. Instead, each design and construction step will reference the appropriate drawings. These drawings can be found in other locations.

Viewing the Drawings for the System

The actual full-color drawing can be found in the Main Version of the Flood Control Design. Many drawings (though not all) are available in the Abridged Version of the Design. Both are available from the author's page at Amazon.

Many drawings are also available for viewing from the author's website: https://markfennellvisionary.com. Drawings may also be provided to the chief engineers and city planners upon request.

Guiding Engineers for Details and Calculations

Guiding the Details of Layout and Locations

The basic concepts of the Advanced Flood Control System can be used in many regions of the world. However, there will specific details which vary depending on your location. We will guide you through these details. This book will help you layout decisions (subjective and engineering based), chose locations (general and specific), and guide you in the important sizing decisions.

It is very important that the engineers place the structures in the ideal locations. Placing each structure in the best location is very important to making this system work as it should. This book will guide you in making the best decisions for placement of each structure.

Calculating for Dimensions and Quantities

The actual construction of any project depends on accurate calculations, for numerous aspects of the system. Yet, the particular dimensions of the parts will vary depending on the needs of the local area.

Therefore, this book guides you through all of the important topics which require specific numbers. You will know what data to collect, what choices to make prior to calculations, and what quantities to calculate.

You will also be given important calculation steps for all pipes, chutes, and containers. Note also that the author has done detailed research on the inadequate calculations for many existing flood management designs. Therefore, if you follow the calculation guidelines in this book, your system will always be able to handle any amount of storm water.

Some calculation details will be left to the engineers. For example, the author is well aware that civil engineers and drilling crews know how to calculate the strength required for force of liquids; and the thickness of materials required to withstand this force. These details are mentioned as reminders, but actual calculations are left to experienced engineers.

You will also be given ranges of values for many items. Minimum dimensions, and likely ranges of dimensions, are given as guidelines throughout this book.

Table of Contents

Introduction

These pages have suggested lists of steps for Design and Instillation of this Flood Control System. These will be grouped as Phases, then with lists of the important steps in each phase.

If you follow these steps, you will be able to construct this Advanced Flood Control System effectively. You will place all items in best locations. You will size all structures and equipment to the dimensions needed. You will also build the entire system in a most efficient manner.

You can of course modify these steps as needed for your location, and based on ideas from the lead engineers. However, these steps are everything you need to properly design and install the entire system.

Installation Process in Brief

The basic process for Installation of this system will be to first dig the large pit of the End Point, yet only to the Main Level at first. This is where everything of the construction will be accessed. It is where all the tunnels and drainage pipes will be built from.

Then we will stabilize with a layer of Concrete. Once this is done, we can build the most important structures of the system. This includes: Building the Flood Chute; Building the Maintenance Tunnel; Installing the Drainage Ring; and Installing the Hydropower System. We can also install the Archimedes Screws, and all of the electrical systems.

When all of this stage is complete, we will dig another earthen ramp, and a much deeper hole. This will give us access to what will be the Absolute Bottom. Now we can bring our trucks in to build: the Flood Storage Container, the Maintenance Room, Filter System, and Primary Pump.

We can then begin to encase the entire thing in the End Point Structure (a box which contains the entire thing), and cover with dirt.

All that remains is to build the Surface Structures, and do some beautification. When these are completed, we will test the system one final time. When the test is successful, the Flood Control System will be officially opened.

All Steps Must Be Done For Each Spoke

Note that ALL of these steps, for ALL of these phases, MUST be done for each Spoke. Remember that there are several spokes in the pattern of the Flood Control System. Each of these spokes is the Flood Chute (and adjacent Maintenance Tunnel), leading to its own End Point. Therefore, each of these Spokes and End Points must be built using the same set of steps.

Ideally, all spokes should be built at the same time. There should be separate crews, and separate machines, at each of the End Points. These crews and equipment will be building the tunnels and pipes for their spoke, at the same time.

At minimum, simultaneous crews MUST be used when boring holes for the Drainage Rings, and when actually installing the Drainage Rings. Everyone must be working on the same Drainage Ring, from different positions, at the same time. This is important so that that every arc of a Drainage Ring is installed with perfect fitting geometry.

Big Picture Overview of Steps

Before getting into the Phases, we will provide a more simplified version: "The Big Picture Steps". These are the main items, in suggested order.

A. Determine Specific Locations and Sizes for all Items.

B. Dig Ramps and Holes at End Points to Reach Main Level.

C. Build Flood Chutes and Maintenance Tunnels.

D. Install Drainage Rings and Entrance Pipes.

E. Finalize the Main Level Area.

F. Dig End Point Deeper to Reach Absolute Bottom.

G. Build Storage Containers, Filters, and Pumps.

H. Finalize the Electrical System and Main Power Cable.

I. Seal the End Point and Build Surface Structures.

J. Beautification Details and Testing the System.

Suggested Phases for Design and Instillation

The following is the list of Suggested Phases for Design and Instillation of the Advanced Flood Control System. Note that each Phase will be subdivided in later sections, into the important steps and related subset topics.

A. Determine Specific Locations and Sizes for all Items

- Phase 1: Select Specific Locations for All Structures

- Phase 2: Determine Exact Sizing for All Structures

B. Dig Ramps and Pit at End Points to Reach Main Level

- Phase 3: Dig the End Point, to Depth of Main Level, with Ramp

- Phase 4: Lay Concrete on the Main Level

C. Build Flood Chutes and Maintenance Tunnels

- Phase 5: Build the Support Ramp for End Section of Flood Chute

- Phase 6: Build Flood Chute, Section by Section, Upward

- Phase 7: Build Underground Point at City Center

- Phase 8: Build Maintenance Tunnel, Section by Section, Downward

- Phase 9: Build Entrance Structures and Shelters
  (Simultaneous with Phase 8)

D. <u>Install Drainage Rings and Entrance Pipes</u>

- Phase 10: Install Drainage Rings – Each Ring at a Time

- Phase 11: Install Entrance Pipes and Grates
  (Can be Simultaneous with Phase 8)

E. <u>Finalize the Main Level Area</u>

- Phase 12: Finish Maintenance Tunnel (Roof and End Point)

- Phase 13: Install the Hydropower System

- Phase 14: Finalize all Power Lines and Main Power Cable

- Phase 15: Install all Maintenance Doors, Lights, Cameras

- Phase 16: Build Elevator to Main Level

- Phase 17: Build Box Walls Around Main Level.

F. <u>Dig End Point Deeper to Reach Absolute Bottom</u>

- Phase 18: Dig Ramp of End Point Longer and Deeper,
  and Clear out Space for the Absolute Bottom

  This will begin the next major construction point

G. <u>Build Storage Containers, Filters, and Pumps</u>

- Phase 19: Build Flood Storage Container

- Phase 20: Build Maintenance Storage Room

- Phase 21: Install Pulley Systems, Panels, Doors, and Ladders

- Phase 22: Build Filtration System

- Phase 23: Install Primary Pump

- Phase 24: Build Archimedes Screws

H. <u>Finalize the Electrical System and Main Power Cable</u>

- Phase 25: Finalize All Pumping Connections and Seals

- Phase 26: Finalize All Electrical Wires for Internal Use

- Phase 27: Construct and Lay Main Power Cable on Tunnel Roof

I. <u>Build Surface Structures and Seal the End Point</u>

- Phase 28: Build the Long-Term Storage for Clean Water

- Phase 29: Construct Elevator Room on Surface

- Phase 30: Build Support Walls and Posts on Storage Room

- Phase 31: Build Walls from Absolute Bottom to Surface

- Phase 32: Lay End Point Roofs to Seal End Point

- Phase 33: Replenish the Dirt and Grass

- Phase 34: Finalize Ventilation Structures on Surface

- Phase 35: Beautification of End Point Meadow and Solar Power

J. <u>Beautification Details and Testing the System</u>

- Phase 36: Build or Finalize Entrance Point Structures

- Phase 37: Test the System

- Phase 38: Create Inspection and Maintenance Schedules

- Phase 39: Project Completed; Grand Opening Ceremonies

Checklists for Each Phase (Overview)

The following sections will expand the above Phases in greater detail. Under each Phase Heading, you will find the most important steps to be done in that Phase.

These details will be important steps, reminders, checklists, and so on. Anything that is very important when completing that particular phase will be listed in bullet form.

Phase 1:

Select Specific Locations for All Structures

The First Phase is to select the specific locations for every structure in the Flood Control System. Engineers and city planners must decide exactly where each item will go.

For illustrations, see Appendixes 1 through 9.

A. <u>Specify the Region and Water Volume for the Flood Control</u> System
- Specify exact geographic region for the System.

B. <u>Make note of all existing structures:</u>
- Existing underground water, gas, and power systems
- River systems (above or below ground)
- Aquifers, caves, and older tunnels

C. <u>Determine Volume of Water through System</u>
- Gather historic data of major floods in region
- Use largest volume water of historic data, plus 30%, for Design Parameters
- See Construction Tips for Additional Details

D. <u>Decide the Number of Flood Chutes and Drainage Rings</u>
- Select the Number of Flood Chutes (usually 4-16)
- Select the Number of Drainage Rings

- These choices will usually depend on the amount of water in the most severe floods, and the locations which need drainage most.

E. <u>Select Specific Locations for all of the following</u>:
- Flood Chutes
- Maintenance Tunnels
- Drainage Rings
- Entrance Grates and Entrance Pipes
- Entrance Points to Maintenance Tunnel
- End Points

Phase 2:
Determine Exact Sizing for All Structures

The sizing of each structure is very important. There are some general sizing guidelines. However, each Flood Control System will have its own specific sizes for each item. The following provide general sizing ranges, and how to do specific calculations, for all aspects of the Flood Control System. For illustrations, see Appendixes #1 through #9.

A. Determine Volume of Flood Water for Region
- Collect the Historic Flood Data
- Use the Maximum Flood Volume of History x 30% as your data
- For Additional Calculations, see Construction Tips section

B. For the Flood Chutes and Drainage Rings, determine the EXACT dimensions for each of the following:
- Flood Chutes Slope Angle
- Flood Chutes Maximum and Minimum Width (Interior)
- Drainage Ring Locations (Geographic Coordinates)
- Drainage Ring Depths (based on Chute Slope)
- Width of Each Flood Chute Section after each Drainage Ring
- Thickness of Flood Chute Walls, Floor, and Ceiling
- Drainage Ring Diameter

C. Determine End Point Location and Depth of Main Level
- Select the approximate location of End Point meadow.

- Based on length of Flood Chute, select the Center of the End Point.
  *The end of the Chute will become the geographic center of the End Point Area.
  *This will be the center of the End Point Design, as well as the Digging and Construction at End Point.

- Based on Slope of Flood Chute, determine the Depth of the end of Flood Chute. *This will also be the Depth of the Main Level*

- To determine total depth and area for End Point, must first learn other measurements (below)

D. For the Structures and Equipment at the End Point, determine the EXACT dimensions for the following:
- Flood Storage Container (Interior and Exterior Dimensions)
- Maintenance Storage Room
- Generator Room
- Filtration System
- Primary Pump

E. Based on the dimensions above, determine the Exact Areas and Depth Needed for End Point:
- Total Area and Exact Depth of Main Level
- Total Area and Exact Depth of Absolute Bottom
- Geographic Center Point of End Point

F. Determine Dimensions for Entrance Points and Shelters
- Diameter and Depth of Entrance Points Stairwells
- Dimensions of Structure for Entrance Points
- Depths and Interior Dimensions of Shelters

G. Determine Dimensions for Entrance Grates and Pipes
- Locations for all Entrance Grates and Pipes
- Diameter and Depth for all Entrance Pipes
- Locations and Dimensions for Entrance Grates

H. Note that for the Maintenance Tunnel:
- The Dimensions should be as specified in construction tips
- The Roof will have supporting beams and a second roof, equal to height of the Flood Chute at that section. (For laying power cable).
- Therefore Dimensions are somewhat pre-determined by other factors.

I. Design and Test on Computer Program
- Design a Computer Program to test system electronically
- Include all Structure and Geological Features
- Use all of the exact data determined above
- Send the maximum volume of water, watch computer results
- Modify values as needed, until satisfied with computer results
- Design phase complete; ready to begin construction.

*These are the most important dimensions to know before beginning actual construction of the Flood Control System. Using these measurements we can begin construction. The dimensions for other items can be determined throughout the construction process.

Phase 3:
Dig the End Point to Depth of Main Level
with Access Ramp

Most of the construction will occur from the Main Level. The building of the Flood Chutes, Maintenance Tunnels, and Drainage Rings, and other items will use the Main Level as the starting point.

Therefore, the first main task in building this system is to dig the End Point down to the Main Level, and prepare it for all the construction to come.

An earthen ramp will lead from the original surface to the Main Level. All trucks and equipment will come down this ramp, and onto the floor of the Main Level. Directly in front will be the Wall of Rocks. The boring machinery will go directly into this wall, and start creating the tunnels.

Therefore, the first important construction step is to simultaneously dig down to the Main Level, and create the Ramp down to that level.

For illustrations, see Appendix #11.

A. Mark the Areas of the End Point
- The End Point is where we begin all construction operations
- Mark the Center Point; this is very end of the Flood Chute
- Rope off the total area of the End Point
- Then Rope off ONLY the Main Level Area

B. Dig a Ramp and Hole to the Main Level (A.11.1)
- Dig the rectangular hole down to the Main Level
- Depth must be exactly to Main Level, plus 5 feet
  - The extra 5 feet for concrete layer
- Area of pit must be at least full area of Main Level, continuous depth from surface to Main Level
- Simultaneously, build Ramp for trucks to reach Main Level
  - Earthen Ramp
  - Gentle Slope, easy access for trucks in and out

C. Save the Soil for Reclamation later
- Set aside the soil, later fill in around the pit
- Save the topsoil, replace after project completed

Phase 4:
Lay Concrete on the Main Level

The Main Level will have a floor of concrete. This concrete provides stability for structures on top of the Main Layer.

A. Place a layer of concrete as Main Level
   - The concrete layer should be 3-5 feet thick
   - Surface Area will be all of the Main Level Dimensions
   - Smooth layer of soil before laying concrete
   - This will become the "floor" of the Main Level

B. Trucks may Drive on the Concrete of Main Layer
   - Trucks and Digging Machines may drive on the concrete
   - Make sure the concrete is fully strengthened before proceeding

Phase 5:
Build the Support Ramp
for End Point Section of Flood Chute

The priority at this stage is to build the Ramp Support for the final section of the Flood Chute. This end of the Flood Chute will actually stick out beyond the earth, into the "room" of the Main Level.

For most of the Flood Chute, the chute will be buried within the earth, and supported by the earth. However, for this section, the chute must be supported by an artificial concrete ramp.

Furthermore, this ramp will be used by the machinery to reach the earth wall, before digging the Flood Chute Tunnels.

Note also that this section of Flood Chute is the widest section of flood chute. It is important to know this when building to right dimensions. In addition, the width allows for easy access of machinery when digging the tunnels.

A. Build Artificial Support for Flood Chute in the Main Level
- Construct a Concrete Ramp Support for Flood Chute
- The Support Ramp will be solid concrete, and very wide
- Length will extend from end of chute to earthen wall
- Width must be minimum that of Flood Chute to be built
- Preferred to have additional 3 ft. width both sides, for stability of walls of this chute
- Slope angle of ramp to match slope of flood chute

B. Digging Machines use Support as Ramp
- Digging Machines will use Ramp to reach Earth Wall, then start boring the hole.
- The chute section will be extremely wide here, therefore ramp will be wide. Plenty of room for vehicles to drive up and down.
- This will continue for the entire Flood Chute Building Process.
- Only when entire Flood Chute constructed will this final section of Flood Chute be enclosed.

C. Dirt Removal Using Ramp

- Dirt removal will also use this ramp.

- All dirt from the Flood Chute digging will be carted down the flood chute hole in small vehicles to the end of the Flood Chute (the End Point). The dirt will then be driven down the ramp support to the Main Level; then across the Main Level, and up the earthen ramp to the surface.

Phase 6:
Build Flood Chute, Section by Section
Upward from End Point

The Flood Chute is the main structure which will carry the water. Therefore each Flood Chute must be built correctly.

The easiest way to build the flood chute is from the bottom up. From the Main Level at the End Point, we start digging holes that will become the Flood Chute. We then create the concrete box which becomes the actual Flood Chute.

It is also best to create the Flood Chute in sections. Each "section" is the length between Drainage Rings.

The primary reason to build in sections is that each section will have different widths. The end section will be the widest. Each section further up will progressively less wide. The reason, as stated earlier, is due to the volume of water. Each Drainage Ring adds more water, and therefore the Chute must be made wider to accommodate. The chute may also be taller after each Ring. When building the chute, therefore, we reverse the order, digging the widest and tallest section first, then progressively smaller dimension tunnel holes up the length.

Another reason to build in sections is to create the actual chute soon after creating the hole. For better stability, it is best to build the concrete box of the actual chute soon after digging the earth. Therefore, we dig a length of hole, then begin creating the concrete chute. Then proceed to the next section, to dig the hole and build the chute. This allows the earth to remain stable, while each section is being constructed.

For illustrations, see Appendix #1, #2 and #11. Most of the pictures in Appendix #1 and #2 demonstrate the Flood Chutes.

Shared Wall, Wider Base and Thinner Top Section (A.11.5)

Note that the Flood Chute will be taller than the Maintenance Tunnel, though both tunnels will share one wall. This means the roof of the Maintenance Tunnel will be placed approximately 3/4 up the height of the Flood Control Chute. To build the roof of the shorter Maintenance Tunnel, therefore, we need to build the shared wall with a wider base (shared wall), then have thinner wall (which is wall for Flood Chute only). This allows a ledge, to set the Roof of the Maintenance Tunnel.

As a bonus, the thicker base of the shared wall will provide extra support for the power of the flood waters through the chute.

A. Dig the Tunnel Holes for the Flood Chutes
- Use machinery to dig the tunnel for the Flood Chute
- Dig the tunnel in sections; between each Drainage Ring
- Remove dirt using same tunnel

B. Constructing Dimensions of Tunnel for Flood Chutes
- Width and Height of section will vary for each section
- Largest near end, progressively smaller going upward

- Dimensions previously calculated
  - Interior Dimensions for Water Volume
  - Floor and Wall Thickness for Proper Strength
  - Floor Thicker than Walls

- Always Wider on Opposite Side of Maintenance Tunnel
  - Maintenance Tunnel constant width
  - Flood Chute width increased opposite wall

- Add a few feet for Maintenance Tunnel side, for the wider base of shared wall.

- Extra Width and Height of tunnel space recommended for workers to place roof.

C. Modify to Go Under Existing Structures
- Sometimes existing structures such as gas pipe lines, power lines, and older tunnels cross will cross the path of the Flood Chute.

- When this occurs, modify the slope of the chute to go under the existing structure. Then increase the slope after the existing structure. Then return to original designed slope angle.

- Note that the flood chute should be built under existing structures rather than over. The Flood Chute will generally be greater dimensions and weight, and therefore should be built under.

D. Build the Concrete Flood Chute, Section by Section

- Install the Concrete Floor for the Chute Section.

- Create the Concrete Walls.
    - Interior dimensions for volume of water
    - Exterior dimensions to include thickness of strength
    - Additional Width to Support *Maintenance* Roof
        (this is to the left of the Flood Chute)

- Add Interior Support Pillars within Chute as needed.
    - Very wide sections of Flood Chute may require 1 or 2 additional interior pillars to support the roof above.

- Install the Flood Chute Roof, to support earth above.

- Seal all Corners for Absolutely Zero Leaks.

E. Shared Wall has Extra Width for Maintenance Tunnel Roof (A.11.5)

- The Maintenance Tunnel and the Flood Chute will share one wall. However their heights will be different.

- To Accommodate the shorter roof (of the Maintenance Tunnel), the Flood Chute wall will have a ledge on which to set the Maintenance Roof.

- This will require wider base wall (by approximately 3 feet) for the "shared" portion of the two tunnels.

- Yet thinner wall, of the Flood Chute wall only, extending upward beyond the height of the Maintenance Tunnel.

- This will allow a ledge of approximately 3 feet width, on which we can later place the roof of the Maintenance Tunnel.

F. <u>Note the Openings for Doors and Ring-Chute Connectors</u>

- The side of chute next to maintenance tunnel will have some openings. These will be for Maintenance Doors and Ring-Chute Connectors. Walls must accommodate these items.

- For Drainage Rings: Holes must be built into the left wall of the Flood Chute, at precise locations for the Drainage Rings.

- Note that the Ring-Chute Connectors, and the Drainage Ring Pipes, will not be installed until later. However the hole for the pipe to enter the flood chute must be ready.

- Doorways for future Maintenance Doors created in the walls, in desired locations.

G. <u>Chute Section Near Main Level Finished Last</u>

- The end section of the flood chute, sitting on the Main Level, will be the last chute section to be completed. All other sections must be completely built first.

- This chute section can be built now, or during the Hydropower Installation (See Phase 13).

- Create the floor, as additional concrete layer on the ramp support, yet a few feet thinner.

- Optional: Add lip at end, for water to pour down easily.

- Install the Turbine; easier to put turbine in its place before walls created.

- Construct the walls of the chute around the turbine, and seal with a roof.

H. Inspect and Test the Chute

- Do a walk-through inspection of entire flood chute.

- Use electro-mechanical devices to measure integrity of walls.

- Begin digging hole under the end of Flood Chute (for test).

- Send small amounts of water down chute, from top to bottom.

- Note any leaks or structural deficiencies, then repair.

- Water will lead to dirt in the hole at very bottom of End Point.

- Note that the full testing of the flood chutes must wait until the entire system is constructed. This stage of testing will catch immediate structural issues and leaks before proceeding.

Phase 7:
Build Underground Starting Point

The Term: "Underground Starting Point"

The "Underground Starting Point" should not be confused with the pit where we dig. The "Starting Point" and "End Point" are reference terms to the flow of the flood water, not the construction process.

The "Starting Point" is the beginning of the Flood Chute; while the "End Point" is the end of the Flood Chute. The "Starting Point" is near the center of the city; the "End Point" is in a rural area far from the city.

For the actual construction process, it is best to build the system in reverse. We begin our construction process at the "End Point". Then we dig upward to the "Starting Point".

Furthermore, the "Underground Starting Point" corresponds to the "Underground End Point" by also being a room underground. However, this room will be smaller, and closer to the surface.

The primary uses of the Underground End Point will be for vehicle turnarounds and access from the street.

For illustrations, see Appendix A.11.2 (a-c)

Uses of the Underground Starting Point

At the very beginning of the Flood Chute and Maintenance Tunnel will be an Underground Starting Point. This is simply the beginning of each of the tunnels. Yet we can use the point for more than that.

The Underground Starting Point will be circular room, where vehicles can turn around. An optional design will include a circular vehicle tunnel, connecting each of the Starting Points together.

We can also install an Entrance Point, with large freight elevators. Alternately, we can install driving ramps from the city streets down into the Starting Point. These can be used for workers to enter the tunnels at this location, and bring vehicles inside.

For illustrations, see Appendix A.11.2 (a-c)

A. Dig the Turn-Around of Starting Point

- The digging machines are already at the upper point of the Flood Chute. Use these to create the Underground Starting Point

- Dig a near circular room, large enough for all vehicles to turn around easily.

B. Dig the Entrance Ramp or Elevator Room

- If the an Entrance Ramp is to be put in, leading from street to the Starting Point, now is the time to create it.

- Use the digging machines to create the ramp up and out, from the Underground Center Point to the street.

- Similarly, if a Freight Elevator is to be installed, use the digging equipment at this time to create the side room for such elevator.

C. Dig the Connection Tunnels for the Underground Starting Points

- Each Underground Staring Point may be connected to the others with an underground tunnel.

- At this time, use the digging machines to dig from one Starting Point to the next.

- Ideally, multiple machines and crew are digging these tunnels at the same time. This will result in all Underground Starting Points and connection tunnels being built at the same time.

D. <u>Begin the Digging of Maintenance Tunnels</u>

- Each digging machine can now begin digging the Maintenance Tunnels.

- Maintenance Tunnels can be dug out from the Starting Point, downward to the End Point.

- The position of the vehicle allows this step easily. There are two choices: where it is now, or return to original location.

    - Digging at current location: The digging machine simply turns in the Starting Point 90 degrees, and begins digging a tunnel parallel to the flood chute.

    - Returning to original location: Alternately, the digging machine can turn 180 degrees, head back to the original area it came from, and dig the Maintenance Tunnel at that location.

- More specific details are in the Next Phase.

36

Phase 8:
Build Maintenance Tunnel
Section by Section, Downward

Main Process

After the Flood Chute is completed, then the Maintenance Tunnel can be built. The Maintenance Tunnel can be built carved from the Starting Point, heading down to the End Point. Digging can also begin from the End Point and work upward. Dirt will be removed in small vehicles, carried down to the End Point.

In general, the Maintenance Tunnel will be easier to build than the Flood Chute, because the structural integrity is not as important as the Flood Chute. However, there are co-existing structures, such as the Drainage Rings and the Main Power Line. Therefore a few modifications must be taken as we build, in order to create the entire set of structures most efficiently.

For illustrations, see Appendix #11 and #2.

Drainage Rings

The Drainage Rings are very important, and therefore the creation of the Drainage Rings, in conjunction with the Maintenance Tunnels, requires some specific planning.

Some areas of the Maintenance Tunnel will be given the concrete walls, while other locations will remain dirt. The floor will also remain dirt. In this way, most of the tunnel will have support, yet we will be able to dig the holes for the Drainage Rings later. Then, after the Drainage Rings are installed, the remaining walls can be put in place, as well as the floor.

Extra Height (A.11.5 a-c)

The Maintenance Tunnel will also have two heights. The first height is the roof of the Maintenance Tunnel. The second height is a second roof, on pillars, above the main roof.

Remember the Main Power Line will be a cable laid on top of the Maintenance Tunnel Roof. The extra space above the Maintenance Tunnel will allow the workers to lay the cable easily. (And to repair it in the future).

End Section of Maintenance Tunnel

The end section will require particular considerations. As with the Flood Chute end section, the Maintenance Tunnel end section will stick out beyond the dirt, and therefore require a support. This support can also be used as ramp for men and vehicles to enter the tunnel.

The end section will also require access to an elevator shaft, which can be built during this time.

Building Tunnels Separately or at Same Time

Note that the Maintenance Tunnel should be built separately from the Flood Chute, due to different heights. Furthermore, building the Flood Chute separately, and in sections as described above, will help maintain stability of the underground operations until completed. It is also wise to focus on the Flood Chute, due to the importance of the Flood Chute structure.

However, if planned effectively, it is possible to build one tunnel, which is very large, then create both tunnels within. There is one shared wall of the Flood Chute and Maintenance Tunnel, thought the Flood Chute will often be taller.

We will continue with the original plan of building each tunnel separately.

A. Build the Supporting Ramp at End Point

- Beyond the earth wall, in the End Point, the Maintenance Tunnel (yet built) will require some support. This is similar to the support ramp for the Flood Chute, just shorter in length.

- This can also serve as a ramp for the trucks to the tunnel (as the tunnel is being built).

- Two crews can work at the same time, one for Support Ramp, the other for Digging. This Support Ramp can be built at this time.

B. Dig the Maintenance Tunnel, Section by Section

- The Digging Machinery will begin at the Starting Point of the Tunnel, where the machinery is currently located.

- Dig the Tunnel according to specifications in this document.
    - Interior Height of 20 ft. minimum
    - Add additional height for second roof.
    - Add width as necessary for ease of construction

- Note this will be earth-rock tunnel only for the moment.

C. End Point Exit

- Digging equipment will eventually poke through earth wall above the End Point, and will drive down the Support Ramp.

- Dirt may be carted out down the tunnel to end of the tunnel, and down the Support Ramp.

- Note that this can be done in reverse order, with digging equipment going up the support ramp, and dirt removed during the process.

D. Carve out Entrance Points and Shelters (See Phase 9)

- Carve out all side rooms, such as the stairwells for Entrance Points, and Temporary Shelters.

- These can be dug at their locations while digging the Maintenance Tunnel.

- Surface Crews will work at the same time, to create the full Entrance Points.

- For further details, see Phase 9.

E. Construct Left Walls for Most of Tunnel

- The Maintenance Tunnel will need support as soon as possible.

- The Maintenance Tunnel will already have a supporting wall on the right, shared with the Flood Chute. However, an additional support must be built on the opposite wall.

- Build Supporting Walls on the left side, for most of the Tunnel.

- However, leave short sections without concrete where the Drainage Rings will be created.

F. Proceed with Side Holes for Drainage Rings (See Phase 10)

- The holes for Drainage Rings may be dug in the side walls as soon as the majority of that side has been reinforced.

- Use a separate machine for digging these holes.
  - Smaller boring machine
  - Computer Driven, with GPS precision
  - Able to make smooth, continuous turns

- Dig holes and install one Drainage Ring at a time

- Multiple crews should work on each arc at same time.

- Dirt can be removed by pushing back to maintenance tunnel, or forward to next flood chute, then carted away.

- Note that only the holes are created. Installation of Drainage Rings must wait until after floor installed.

- For further details on this step, see Phase 10

G. Lay the Floor for Maintenance Tunnel

- Lay the concrete floor as soon as the holes for the drainage ring holes have been dug out.

- Lay the entire floor of Maintenance Tunnel before installing any Drainage Ring pipe.

H. Install Drainage Rings in the Arc (See Phase 10)

- After the Maintenance Floor is laid, the Drainage Rings can be installed in the holes.

- Drainage Rings must be installed arc by arc, ring by ring.

- Ideally, multiple crews will install each arc of one ring at the same time.

- Robotic machines can pull the metal pipes through the earth tunnel.

- Robot on wheels will grab section of pipe, pull forward, then drive back to get next section of pipe. These robots will be driven remote control by workers in the Maintenance Tunnel.

- When all the Drainage Pipe Sections are approximately in their place, then men can walk through the pipes, and adjust their precise location.

- Drainage Ring Sections will then be sealed water proof.

- For further details on this step, see Phase 10

I. Install Drainage Rings in Ring-Chute Connector

- The Ring-Chute Connector can be put in place after the floor of Maintenance Tunnel has been created.

- The Ring-Chute Connector is a concrete block, almost the full width full of the Maintenance Tunnel, with two parallel holes.

- Place the Ring-Chute Connector so that the one hole matches the piping hole of the Flood Chute Wall.

- The Drainage Ring Pipe will be inserted into this Connector.

- The Pipes are sealed at the left wall, and a gentle ramp is built over the Ring-Chute Connector.

- For further details, see Phase 10

J. Finish the Maintenance Tunnel (See Phase 12)

- After the Drainage Rings are put into place, the Maintenance Tunnel can be finished.
- Construct wall sections around each Drainage Ring.
- Install the roof of the Maintenance Tunnel
- Construct the pillars (and second roof if desired) above main roof
- Build the full Maintenance Tunnel Box at End Point
- For further details, see Phase 12

Phase 9: Build Entrance Structures and Shelters
(Simultaneous with Phase 8)

The Entrance Structures and Shelters can be built at any time. These will be constructed primarily from the surface downward, and by a separate crew. However, it can be convenient to build these structures at the same time as digging the Maintenance Tunnel.

For illustrations see Appendix #9.

A. Create Stairwell
- From surface, dig stairwell hole
- Proper depth to reach Tunnel at coordinates
- Proper diameter for staircase
- Build supporting walls from ground to surface
- Install Stairwell
- Ideal to build Stairwell taller than surface (see below)
- Ensure doorway to Maintenance Tunnel is stable

B. Create Shelter Rooms
- Working with Tunnel Crew, create shelter rooms where desired
- Reinforce rooms with concrete walls and roof.

C. Build Earthen Mound for Rain to Flow Off
- Build earthen mound surrounding hole (water flows off)
- Height will match top of staircase

D. Build Surface Structure for Entrance Point
- Lay the concrete base surrounding the stairwell hole. This will become the room of the structure.
- Build guard rails around the stairwell, and rails on the stairs.
- Construct walls of stairwell, to sufficient height, with one outer door.
- Install the angled roofs
    - Material very durable; metal is ideal.
    - Slope Angle and Direction to best for Solar Power
    - Screened gutter system to channel rainwater
- Construct a drinking water system, for roof rain water, with small filter unit and large container

E. Install the Electrical Systems of the Entrance Structure
- Install the batteries for solar power storage in Tunnel below
- Install Solar Power arrays on two sides of roof
- Install radio wave antennas and surveillance cameras
- Connect all the wiring to function properly

F. Beautification and Safety
- Review the entire structure and stairwell for safety.
- Do architectural beautification and landscaping.

Phase 10: Install Drainage Rings
Each Ring at a Time

Main Process of Installing Drainage Rings

Drainage Rings are almost as essential to the Flood Management system as the Flood Chutes. Furthermore, these Rings will be connected in arc segments. Therefore, the Drainage Rings must be built with precision.

Holes will be drilled in arc segments, between spokes of the flood system (specifically from the Maintenance Tunnel of one spoke, to the Flood Chute of the next spoke). Pipes will then be inserted, in short curved pieces, and sealed together.

At the Maintenance Tunnels, the final pipe will be connected and inserted into the Ring-Chute Connector.

In the Flood Chute, the pipes will extend into the chute. Both sides of the chute will have pipes leading into the chute, allowing water to pour into the Flood Chute easily.

For illustrations, see Appendix #1, #2, and #11.

Construct Each Drainage Ring Before Attempting the Next

It is best to build the Drainage Rings one at a time. Each Ring should be built in its entirety, using multiple crews for each arc. When the Drainage Ring is complete, then proceed to the next one.

Precision Drilling and Inserting

Holes must be aligned with exact depth at each location. Furthermore, the holes must be dug with precise geometry such that each arc connects tunnels perfectly. Therefore, special digging machines will be used which have computer maps and GPS systems.

Each Drainage Ring will be built in arc segments. Each arc will connect the Maintenance Tunnel to the next Flood Chute. (Not the adjacent flood chute, but the flood chute of the next spoke).

Furthermore, it is best if multiple crews work on the arcs simultaneously. This will ensure that the holes are drilled and pipes are inserted at the precise geometry all the way around.

Robots and Humans

The initial drilling will be done by robotic machines, with computer spatial awareness and GPS. Pipes will be pulled in by robotic vehicles. Both of these can be remotely driven by humans. Finally, workers themselves will enter the pipe to do final placement adjustments and permanent seals.

A. Drilling the Holes for Drainage Rings

- Holes for Drainage Rings will be near the bottom of the Maintenance Tunnel.

- Use GPS and similar systems to get precise positions to begin digging the hole.

- The holes for Drainage Rings may be dug in the side walls as soon as the majority of that side has been reinforced.

- Use a separate machine for digging these holes.
  - Smaller boring machine
  - Computer Driven, with GPS precision
  - Able to make smooth, continuous turns

- Dig holes and install one Drainage Ring at a time

- Multiple crews should drill holes for each of the arcs at same time and communicate with radio equipment to ensure that the holes are aligned properly.

- Ring Tunnels should be larger than diameter of pipe, for easier insertion and assembly. Extra space is later filled with dirt.

B. Dirt Removal

- Dirt can be removed by pushing to either end of the Drainage Ring tunnel.

- This dirt will pile in either the Maintenance Tunnel of original spoke, or the Flood Chute of the next spoke.

- Cart this dirt to the End Point, and up to the surface.

C. Material of Drainage Rings

- Structure of Drainage Rings should be Steel for strength
- Interior lining should be Stainless Steel or Copper
  - To resist corrosion
  - And possible bacteria reduction

D. Inserting Drainage Rings

- Drainage Rings will be inserted in small sections, for each arc.

- Each section of pipe will be short curved pieces.

- Sections of Drainage Rings can be inserted from either the Maintenance Tunnel side, or the Flood Chute Side.

- Sections will be numbered, to know the correct order to put in
  - This is important due to the curvature of the Ring Tunnel

  - Some sections of Ring might be modified by design, based on the surface rivers and buildings at those coordinates. Proper placement of specific ring sections is important here.

- The tops of many sections will have pre-drilled holes. The entrance pipes, carrying water from the surface to the drainage rings, will be inserted into these holes.

- Robots will pull the Pipe Sections into place
  - Vehicle robot, remote controlled
  - Smooth refined turning ability
  - 4 or more clasping arms to hold pipe
  - Pulls forward, sets in approximate location
  - Robot will drive through placed sections to get next one
  - Exact location and fitting of section will be done by men

- In Flood Chute, pipe will extend at least 2 feet into the Chute
  - This allows water to fall into Flood Chute easily

- In Maintenance Tunnel, pipe extends 6-12 inches in Tunnel
  - This allows for Chute-Ring connection installed

E. Final Placement and Sealing of Drainage Rings

- Men will enter the Drainage Pipe Tunnel, walk to each section, then pull each one manually into exact position.

- Men will then fit the sections together precisely, and seal the connections.

- Sections should be sealed with liquid metal or plastic which hardens.

- Additional sealant can be provided from metal rims soldered on the interior.

F. Use of Multiple Crews in the Process

- It is strongly recommended that multiple crews are used in this process.

- Each crew will drill the tunnel and insert the pipes for their arc of the drainage ring.

- Each crew will also communicate with other crews, and monitor their activities, in order for the geometry of the Drainage Ring to be precise.

- Robots for drilling and installing may be shared when needed.

G. Ring-Chute Connector (A.2.12)

- The Ring-Chute Connector will provide extra stability of the pipeline when it crosses the Maintenance Tunnel.

- The Ring-Chute Connector is ideally built as separate pieces, then fitted together.

- The Parts of the Ring-Chute Connector are:
    - "First Connector" = placed against Flood Chute
    - "Pipe #1" = inserted into First Connector and Chute
    - "Main Connector" = most of width of Tunnel
    - "Pipe #2" = inserted into Main Connector
    - "Pipe #3" = pipe from Drainage Ring

- Each part will be placed in the order above, then the pipe segments are welded together. Specific process in next bullets.

- Place the First Connector against the wall of Flood Chute, so that the holes line up.

- Insert pipe #1. Make sure the pipe extends into the Flood Chute at least 2 feet. The other end of the pipe will extend from the Connector 1-2 feet.

- Place the Main Connector across the width of the Tunnel, pushed to approximately 1-2 feet from the extended pipe of the First Connector.

- Insert Pipe #2 through the Main Connector. This Pipe #2 will be pushed against pipe #1, and can be welded together.

- On the other side of Main Connector, Pipe #2 also extends, touching the Drainage Ring Wall Pipe. These can now be welded together.

- The Drainage Ring Section across the Maintenance Tunnel is now complete.

H. Man-Holes Aligned to Access Interior of Drainage Ring

- Note that there are man-holes (and covers) in the tops of both the Connectors and the Pipes in this Tunnel. This will allow access for maintenance workers to access the interior of the Drainage Ring, and travel to any region of the Drainage Ring.

- Man-holes must be aligned when inserting pipes into the connectors. The Man-Hole covers must also be fitted and secured at this time.

- The access hole in the ramp (see below) must have a secure cover so that people can walk over easily. The access hole in the pipes must have a cover to prevent water from spraying out. However, the access hole in the concrete connector may or may not need a cover.

H. Ramp over Ring-Chute (A.2.11)

- Build a gentle ramp over the Ring-Chute Connectors.
- Gentle slope up and over for people to walk across easily.
- Wide as the maintenance tunnel
- Made of wood or concrete
- There must be an access hole in the top of ramp, for workers to enter the pipes below. Secure cover is required.

I. Dirt Fill and Concrete Wall

- Add filler soil around the Drainage Ring in the arc
- Construct concrete wall around drill point in Tunnel
- Finalize any seals or support structures related to Drainage Rings

J. <u>Quick Testing of Drainage Ring</u>

- The Drainage Ring System should be given a quick test

- Each Drainage Ring checked and tested separately

- Mandatory manual walk-through for entire Ring
    - Inspect, Adjust, and Repair as needed
    - Multiple crews can inspect their arcs simultaneously

- Quick water test: Pump water from one Flood Chute to another through the Drainage Pipes
    - Water hoses brought up Flood Chute from End Point
    - Insert into Drainage Ring
    - Watch water flow through arc, through pipes in tunnel, and into the next Flood Chute.
    - Do a second walk-through, to inspect for any leaks of the pipes in the arc.
    - Repeat for all arcs in the Ring

K. <u>Repeat Processes for Next Drainage Ring</u>

- If Drainage Ring inspection and test is acceptable, then this Drainage Ring is complete.

- Proceed to the next Drainage Ring location, and repeat process

Phase 11
Install Entrance Pipes and Grates
(Can be Simultaneous with Phase 8)

Basic Concept
The flood waters will initially enter the Flood Control System through the Entrance Grates and Entrance Pipes. Water will flow from the streets to the Entrance Grates, down into the Entrance Pipes, and directly into the Drainage Rings below.

We can install as many Entrance Grates on the surface as we want, as many Entrance Pipes to the Drainage Rings. However, we must plan exact locations for each before we begin. Furthermore, the actual installation of these pipes can be done simultaneously with the installation of the Drainage Rings.

For illustrations, see Appendix #1, #2, and #11.

Determining Precise Locations for Entrance Pipes
The precise locations of the Entrance Pipes must be known, because there will be matching holes in the pipe sections of the Drainage Rings. Engineers need to know exactly which sections of the Drainage Ring Pipes to put the matching holes, and then be able to place the section in the exact position.

Therefore, the precise geographic coordinates for each Entrance pipe must be determined prior to Drainage Ring installation. The matching holes in the Drainage Pipe Sections will be pre-drilled, then each section will be placed into a specific location.

Number and Size of Entrance Pipes
The most important desire of this Flood Control System is fast drainage of the flood waters. The system on the surface must be designed such that the heavy rains are quickly drained from the streets. Therefore the we want many Entrance Pipes, and we want these pipes to be of large diameter.
- Number of Entrance Pipes = As many as possible
- Minimum Distance between Entrances = 500 feet
- Diameter of Entrance Pipes = As wide as practical
- Minimum Diameter of Pipes = 3 feet

Installation Process (A.11.3)

The instillation process actually begins with the Drainage Rings. The Drainage Rings are installed with pre-drilled matching holes for the Entrance Pipes. Then, on the surface, the hole for the entrance pipe will be drilled, and the entrance pipe inserted deep into the earth.

Men in the Drainage Ring will then perfectly fit the Entrance Pipe to the Drainage Ring. Note that the pipes will be allowed to stick through the Drainage Ring a few inches. The men will then seal the two pipes together.

A. Preparing Drainage Ring Holes and Signaling to Surface
   - Pre-Drill matching holes for Drainage Rings Sections
   - Insert Drainage Ring Sections according to design plan
   - Attach magnetic radio signal device to pipe near hole
   - Signaling device will assist surface drillers in precision digging

B. Drilling Holes and Installing Entrance Pipes (A.11.3)

   - On surface, use GPS and signaling device to locate precise drilling location.

   - Drill the hole for the Entrance Pipe, to depth of Drainage Ring Tunnel. (Tunnel is wider than Pipe until filled later)

   - Insert Entrance Pipe from surface, with guidance from men at Drainage Ring.

   - Men at Drainage Ring fit Entrance Pipe to Drainage Ring precisely, and seal the two pipes together.

C. Creating Entrance Grates (A.11.4)

   - Build the concrete base surrounding each Entrance Pipe
   - Cover the Entrance with a Grate, to prevent debris inside
   - Indent the Entrance compared to surrounding area, for natural flow of water into the Entrance

Phase 12:
Finish Maintenance Tunnel (Roof, End Point)

The Maintenance Tunnel requires additional construction before it can be complete. These construction steps can only be done after the Drainage Rings and Entrance Pipes have been installed.

For illustrations see Appendix #7 and #11.

1. Walls around Drainage Rings

Up to this point, the Drainage Ring enters the Maintenance Tunnel through a wall of dirt. We can now fill the space around the pipes with filler soil. Then construct the concrete wall surrounding the Drainage Ring Extension.

2. Build the Roof Systems

At this time, the Maintenance Tunnel has floor and two walls, but no roof system. We can now install the roof on top of the walls of the Maintenance Tunnel.

Furthermore, only one of these walls (the shared wall with the Flood Chute) is actually supporting the earthen ceiling. We will also install pillars on top of the Maintenance Tunnel roof, which will provide additional support of the earth ceiling. In addition, we will use this space above the roof to lay the Power Cable.

3. Construct Maintenance Tunnel at End Point

At this moment, there is no Maintenance Tunnel at the End Point. There is just a support for this future Maintenance Tunnel. We have also been using this support as a ramp for equipment to go in and out of the actual Maintenance Tunnel. However, and actual maintenance tunnel needs to be completed.

This will be simple box construction of the maintenance tunnel, leading down the support ramp, to the End Point. In the final version, men will walk through the maintenance tunnel in the earth, then at the End Point, and out the final door.

Note that we could just provide a single wall and door at the earth, then walk down the supporting ramp. This is an option. However, creating the extension of this Maintenance Tunnel into the End Point provides extra safety measures, regarding water flow and geological shifts. It also allows for additional modifications related to flood management in the future.

A. Finish the Walls around Drainage Rings

- Add filler dirt around the Drainage Rings in the arcs as desired

- Build walls surrounding the extension pipes of the Drainage Rings in the Maintenance Room

B. Install Main Roof above Maintenance Tunnel (A.11.5c)

- Install the Main Roof of the Maintenance Tunnel.

- For the side of the roof next to Flood Chute, the roof will sit on the ledge of the shared wall. (See above in Flood Chute Phase).

- Build roof access doors and interior ladders to the Main Roof
  - Maintenance Tunnel Roof accessed from inside
  - Every 1/2 mile minimum.

- Ideally, shape the center line of roof with ditch, where the future Power Line can nest easily.

C. Install Secondary Roof with Pillars (A.7.3 and A.11.6)

- Create a Secondary Roof by installing pillars on the Main Roof.

- Pillars primarily needed on left side (opposite side of Flood Chute wall) where there is no support.

- Pillars will be placed evenly along the left side of the main roof, reaching from the main roof to the earth above.

- Center of Maintenance Tunnel Roof must remain unobstructed, for the future laying of Power Cable

D. Build Maintenance Tunnel End #1
- The end of the Maintenance Tunnel at the earth will require final construction and safety measures.
- Where earth and tunnel meets end point, build a strong wall and sturdy door. This will be the final wall and door before reaching entering the end point.

E. Build Maintenance Tunnel Section at End Point

- It is best to not only have an exit wall and door, but a continuation of the maintenance tunnel, at the End Point.

- At the existing support ramp, construct the floor, walls, and ceiling of the final section of maintenance tunnel.

- Continue with the Maintenance Tunnel to the floor of the End Point.

- The final exit will not need a door, simply the Tunnel Opening.

The Maintenance Tunnel has now been completed.

Phase 13:
Install the Hydropower System

The Hydropower System will be relatively simple to install, compared to the previous structures, as there are fewer objects to put into place. The turbine will require the greatest focus. All the other equipment is standard technology. For illustrations, see Appendix #3 and #4. Most of the pictures in Appendix #3 and #4 are dedicated to Hydropower.

A. Install the Turbine (A.3.7, A.4.1 to A.4.5)
  - Bring the Turbines up through the Flood Chute
  - Place at proper locations, and proper orientation
  - Insert the Axis which will connect to Generator
  - Connect Turbine securely to both walls with steady bar

B. Install the Generators and Transformers (A.3.5 to A.3.8)
  - Lay the concrete floor for the Generator Room
  - Install the Generators
  - Install the Transformers
  - Use Higher Voltage Transformers for Pumping Systems
  - Use Normal Voltage Transformers for Main Power Line
  - Install storage batteries as desired.

C. Construct the Generator Room
  - Construct the Generator Room around the Equipment.

  - Easiest to build floor and install equipment first, then walls around the equipment.

  - Generator Room must be large enough for air circulation, with plenty of space above and around all equipment.

  - Generator Room must also have window openings on each side, such that air can flow in and out of the Generator Room easily.

  - Large doorway will be put in place for inspections and future equipment replacement.

  - Connect all the main power lines from between Turbines, Generators, and Transformers.

59

Phase 14:
Finalize all Power Lines and Main Power Cable

There will be several power lines, each going to different locations. The exact number of power lines depends on the number of turbines and generators. Similarly, the exact configuration of power lines to their destinations will depend on the number of generators and transformers.

The different power line uses include: Primary Pumps; Archimedes Conveyance Screws; Elevators; Lights and Doors in Maintenance Tunnel; and the Main Power Line to the Public.

For illustrations, see Appendix #3 and #4.

A. Connect all Power Lines for Internal Use
  - Primary Pumps
  - Archimedes Conveyance Screws
  - Elevators
  - Ceiling in Maintenance Tunnel

B. Lay the Power Cable for Main Power Line

  - Install the Power lines into the Underground Cable.

  - Lead the Power Cable from the Transformer to the Maintenance Tunnel, and onto the Roof.

  - Lay the Power line continuously on the roof of the Maintenance Tunnel all the way up.

  - The Power Line will eventually be raised up to the surface and to the public. This can be done now or any time in the future.

C. Maintenance Tunnel Wiring

  - Install wiring along the entire ceiling of Maintenance Tunnel
  - This can be done all at once, or in sections when wiring doors and lights.

Phase 15: Install all Maintenance Doors, Lights, Cameras

The Maintenance Tunnel will have Maintenance Doors, Surveillance Camera, and Lighting throughout the tunnel. The power for these items will come from either a) solar power and batteries, or b) hydropower.

The wiring and installation for these electronic items will be simple to install. However, there are some practical arrangements to consider, which are discussed here.

Any of these wiring projects can installed at any time after Phase 14 above. We can install these now, or any time during the End Point construction period.

In addition, the End Point itself will need power, primarily for lights. This power will be provided by a diesel generator, perhaps the same one that powers the elevator. However, we cannot install this until after the End Point is completed. For illustrations, see Appendix #4 and #9.

A. Solar Power, Batteries, and Entrance Point (A.9.2)

- Every Entrance Point to the Maintenance Tunnel will have Solar Power and Batteries.

- If these haven't been installed, now is the time to install them.

- Solar cells are installed on the roofs of the entrance structure

- Batteries will sit in the Maintenance Tunnel below.

- Best to install a platform, near to the ceiling, for batteries.
    - Nearer to the ceiling, where power will be needed

- Connect the wires or the solar cell to batteries, then batteries to doors and lights.

- Automatic "on" when surface door is open. Closed surface door disconnects power from battery.

- This power will run for a short distance on both sides of the Entrance Point, operating lights and doors nearby.

- Areas between the Entrance Points will be powered by the dedicated power line from the hydropower batteries.

B. Line the Ceiling with Dedicated Power Line

- There will be a dedicated power line from the hydropower system to the Maintenance Tunnel.

- Storage Batteries in the Generator Room will store electrical power generated from the Hydropower System.

- These batteries will then be wired to the ceiling of the maintenance tunnel.

- Note that we don't need batteries for pumps, as these will be operational in real time as the Hydropower is created.

- The power line will then be lined along the ceiling of the Maintenance Tunnel, for the entire length.

- Branch extensions will lead to each light, camera, and door along the way.

C. Maintenance Doors: Electronic Locks

- Maintenance doors will be installed along the length of the tunnel. These doors provide access to the Flood Chute, for periodic inspection and maintenance.

- Maintenance Doors will be thick and sturdy, similar to bank safe doors. This is to ensure no waters will enter the Maintenance Tunnel.

- Doors are locked electronically, with a security code.

- For safety reasons, the door will remain unlocked until the security code is entered again.

- Power for these electronic doors will come from storage batteries. The initial power will come from: a) solar power, or b) hydropower, depending on the location of the doors.

D. Maintenance Doors: Placement and Specific Powering

- The locations for Maintenance Doors are pre-determined.

- There are already doorways in the shared wall between Flood Chute and Maintenance Tunnel. Doors simply must be installed.

- Maintenance doors should be built near every Entrance Point. This will provide convenient access for workers and equipment.

- Maintenance doors near the Entrance Points will be powered by local storage batteries (which were originally loaded by solar cells).

- Furthermore, if the Maintenance is being done during sunny days, then the power will be continuously replenished throughout the day.

- Maintenance doors will also be installed at locations between the Entrance Points. These will be powered by the storage batteries at the End Point, which were originally loaded from the Hydropower.

E. Lighting

- Lights will be installed along the entire length of the Maintenance Tunnel.

- Power for the lights will be from storage batteries, loaded either by solar cells or hydropower.

- Lights near the Entrance Points will use the batteries under the Entrance Point.

- Lights in all other areas will be connected to the dedicated power line for the Maintenance tunnel, which is powered by hydro-power batteries at the End Point.

- Any lights can be used, as long as they are:
  - Bright enough for clear visibility
  - Energy efficient (less use of battery power)

F. Surveillance Cameras, Antenna, and Dedicated Battery

- Surveillance cameras can be installed at various points along the Flood Chute to monitor the operations and structure.

- Ideal locations are near the Entrance Points and Drainage Rings.
    - Near Entrance Points provide easy use to solar power.
    - Near Drainage Rings will monitor the flow and any issues caused by the intensity of the water.

- The surveillance camera will be placed at the ceilings of the Flood Chute, secured with very sturdy bracing.

- Electrical power and visual transmission will be sent in wires connected to the surface.
    - Tiny holes in the flood chute wall will allow wires to poke through.
    - Power line is connected to dedicated battery.
    - Transmission is connected to Antenna on Entrance Structure.

- The power to the surveillance camera must be continuous, in all circumstances. Therefore the camera must have its own dedicated power line. The camera should also have its own battery.

- Surveillance cameras near the Entrance Points will be powered by solar cells on the Entrance Structure, then to dedicated battery.

- There will be a dedicated battery for the surveillance camera and antenna.
    - This battery will be independent of all other batteries.
    - This battery will always be "on", powering camera and antenna during storms.
    - Ideal location is in Entrance Structure itself, in a secure box, with holes for warm air to flow out.

- For surveillance cameras near the Drainage Rings, place the cameras just before Drainage Rings enter the Flood Chute.
    - Provides visual of water coming into chute.
    - Will not be damaged by power for water rushing in.

- Power for surveillance cameras near the Drainage Rings will come from Hydropower system. This will use same power line as for the light and doors in maintenance tunnel.

- Note that the hydropower will be working when the Drainage Rings are pouring in water. Therefore power will always be provided for these cameras from hydropower during storms.

- Transmission of the visuals from the camera will be sent in a cable alongside the Main Power Cable, on the roof of the Maintenance Tunnel.

- This transmission can be sent as cable all the way up on the roof, to the managers at the city center.
  - Other transmissions from additional cameras can join this cable, and together reach the managers.

- Alternately, the visuals can be wired to nearest Entrance Point, and sent out through the Antenna.
  - Separate Antennas, or separate Antennas on one stand, may be required for transmission of each of the cameras

G. Power for the Main Level Lighting from Surface Generators

- The Main Level of the End Point will also require power. In general, this area will require power lines for:
    - Lights
    - Mechanical Maintenance Equipment
    - Mechanical Ventilation Systems

- Lights will be placed along the ceiling of the Main Level, and along the walls of the Absolute Bottom Level.

- Power extensions (essentially a place to "plug in" repair equipment) may be led from the surface to the Main Level, and placed along any of the walls.

- Note that all of this power will come from diesel generators on the surface, and not from within the End Point itself.

- Furthermore, the lights and other power extensions cannot be put into place until the roof is put into place.

H. Other Electrical Items

- Any other electrical items in the Maintenance Tunnel can also be installed at this time.

- One such option is to operate the maintenance doors remotely.
    - All maintenance doors can be open at the same time.
    - Or, each door may be opened individually.
    - Wire the doors appropriately, then connect to nearby antennas

Phase 16:
Build Elevator to Main Level

An elevator is installed from the Surface to the Main Level. This elevator will carry crew and equipment to the interior of the end point, for all future inspections and maintenance.

The Elevator can be installed at any time after the first few sections of Maintenance Tunnel are complete. If the elevator has not been completed yet, be sure to do so at this time.

Notice that the elevator will not be a necessity until after the end point is completed and the ground is filled in. However, the elevator can be installed and used when desired during the End Point construction period.

The power for the elevator will come from two sources: diesel generator on the surface, and wiring from the hydropower system. The primary power will be the diesel generator on the surface next to the elevator room.

The elevator is a freight elevator, and therefore must be wide enough for all possible large equipment replacement parts. This includes filters, turbine blades, etc.

The elevator itself can be built now. However, the elevator room on the surface does not need to be completely built or refined until the completion of the project.

For illustrations, see Appendix #4, #6, and #8.

A. Install Elevator

- The elevator should be placed next to the Main Level and/or next to the Maintenance Tunnel. Doors open directly to Main Level or Maintenance Tunnel.

- Dig a hole for the shaft, and install the elevator.

- Install the diesel generator to power the elevator.

- The Elevator Room should be completed later during the beautification phase.

Phase 17:
Build Structural Boundary Walls
Around Main Level

The Structural Boundary

The End Point will ultimately have walls and ceiling built up around it, with landscaping on the top. These walls will protect the entire End Point.

We have referred to this elsewhere as the Structural Boundary of the End Point. This structure will surround the entire End Point, as a very large room.

Note that the Structural Boundary will be deeper than the surface by many feet. After the roof has been placed, there will be dirt above between the Underground End Point and the Original Surface.

For illustrations, see Appendix #7 and #11.

Building Structural Boundary in Two Stages

This Structural Boundary will be built in two parts: Over the Main Level, then over the Absolute Bottom section.

It is easier to build the Structural Boundary for the Main Level at this time. This is because after this phase we will dig a completely new hole, and the Main Level will be more difficult to access.

The Floor and Walls of the Structural Boundary at Main Level

There will be three walls at this level: right, left, and back. There is no front wall at this level. Note that these walls are separate walls, beyond the Flood Chute and other structures.

The floor of the Main Level must therefore be built much wider and longer than all the structures. The right side of the floor, for example, will be much to the right of the Flood Chute. The left side of the floor will be much to the left of the Generator Room. These details should have all been in the dimensions in the original calculation, and in the original construction, of the Floor of the Main Level.

The Right wall will be far beyond the right side of the Flood Chute. It will be a strong sturdy wall, build from the Main Level floor to the pre-determined height. The Left wall will be identical on the other side.

The Back wall will require important designs. The back wall is the barrier between the dirt wall and the Main Level. There will be three holes in this wall: a) for Flood Chute, b) for Maintenance Tunnel, and c) for door to the Elevator. The walls will be built around these structures. The structures (already completed) will pass through the holes in this back wall, to the dirt beyond. (See figure A.7.4 and A.7.5)

A. Build the Back Wall, with Proper Holes for Tunnel and Chute

- Build the back wall in sections, with decent sized holes for Maintenance Tunnel, Flood Chute, and Door to Elevator.

- Back wall must be thick and sturdy, to support the roof that will be placed on top, and to hold the earth wall in place.

B. Build the Right and Left Wall

- Construct the Right Wall, beyond the width of the Flood Chute
- Construct the Left Wall, equal dimensions to the Right Wall

C. Install the Roof

- The Roof of the Structural Boundary will protect the entire End Point for the surface above.

- It is best to lay the roof of the Main Level section at this time.

- Create two holes in the roof (6 inches diameter) for electrical wiring.
  - Hole #1 for power line to ceiling lights
  - Hole #2 for power extension line (for maintenance equipment)

- Add additional holes for ventilation pipes as desired.

- Lay the Roof of the Structural Boundary.

- This room may also be installed later if desired.

Phase 18:
Dig Ramp of End Point Longer and Deeper, and Clear out Space for the Absolute Bottom

The Next Location and Major Step

The next major phase is to dig a deeper hole, to the Absolute Bottom. This step is in fact a Major Step, a Major Phase, in the Construction Process.

Earlier we dug the pit and access ramp to the Main Level of the End Point. From there, we were able to build everything at that level. This included: Flood Chute, Maintenance Tunnel, Drainage Rings, Hydropower Systems, and Electrical Systems.

Now that those are completed, we can move to the next location. We can leave the Main Level, and dig deeper, down to the Absolute Ground. At this location we will construct: Flood Storage Container, Maintenance Storage Room, Filter System, and Primary Pump.

For illustrations, see Appendix #11, particularly A.11.7.

Depth and Area of the Construction of Absolute Bottom

We will repeat the general process of digging the pit, and a gradual ramp for trucks to enter the pit.

The depth will be from the original surface, down to the level of "Absolute Bottom".

The area will be large enough for all objects which are built on the Absolute Level. This includes: Flood Storage Container, Maintenance Storage Room, Filter System, and Primary Pump.

Additional space is needed in front of the filter and pump for people to walk easily in front of the objects (after the objects are put into place).

Beyond this amount of area, dig as much area as needed, at that depth, to easily perform construction operations.

Gradual Ramp

A gradual ramp down into the pit will be constructed at the same time. Trucks, equipment, and workers will reach the Absolute Bottom from this ramp. Note that after this location is complete, the ramp area will be filled with dirt, again to the original surface level.

Absolute Bottom vs. Flood Storage Container Bottom

It is important to notice the difference between Absolute Bottom and the Bottom of the Flood Storage Container.

The "Absolute Bottom" is exactly that. This is the absolute bottom ground for any person or object. Nothing is below this depth.

However the Flood Storage Container Bottom is much higher than the Absolute Bottom. The Flood Storage Container will sit on a thick layer of earth, several feet high. This is important, because the Filter System will be deeper than the Flood Storage Container.

Therefore, the Flood Storage Container will be built on this thick layer of dirt, several feet above the Absolute Ground. The Filter System will then be built on Absolute Ground. Water will flow from the bottom of the Flood Container into the top of the Filter System.

Therefore understanding the difference in levels is important to the construction and operation of the system.

A. Measure out the Surface Area Needed
- Area for all structures at Absolute Bottom
- Plus area for workers and equipment during construction
- Plus area for ramp to Absolute Bottom

B. Dig the Pit and Ramp
- Dig the pit to the Depth to Absolute Bottom, and Area required.
- Build the earth ramp from surface to End Point.

Phase 19:
Build Flood Storage Container

It is now time to build the Flood Storage Container. Ultimately, all the flood waters will be sent into this Container. Therefore the container must be built large enough and sturdy enough for the purpose.

Calculations for the dimensions of this container are described elsewhere in the proposal. Essentially, all of the volume of the water through the Drainage Rings and Flood Chute, will be contained in this structure.

Also remember that the container will sit on top of a tall base, either earth or concrete, to provide sufficient height above the filter system.

For illustrations see Appendix Pictures: A.7.3, A.7.5, A.11.9, A.11.10, and A.11.12. Also see pictures such as A.6.7.

A. Build Earth or Concrete Base for Flood Storage Container

- Build a Base for the Flood Container to sit on.
- Base can be earth or concrete.
- If Base is concrete, allow full time for concrete to seal.

- Area must be minimum of Area of Flood Container
- Add Area for extra support of Flood Container.
- Height will be from Absolute Bottom to top of the Filter Unit.

B. Create the Flood Storage Container

- Construct the Flood Storage Container on top of the Base.
- Concrete floor and walls.
- Height from Base to Main Level (just under flood chute).
- Important to give enough time for concrete to seal completely.

C. Access Door and Filter Hole

- The Flood Storage Unit will be constructed with two openings:
    - Access Door on the Left Wall.
    - Opening to Filter System on Bottom Left.
- A large doorway will be created on the Left Wall, near the floor.
    - A sturdy door will be constructed here.
    - This will lead to the Maintenance Room.

- A smaller opening is created on bottom left for filter system.
    - Hole approximately 1 ft. diameter.
    - Piping to Filter Unit will be inserted here.
    - Allow a few inches above floor for stability of wall.

C. Interior of Flood Storage Container
- Line the interior of the Flood Container with Stainless Steel.
    - Make sure concrete is fully set before proceeding.
    - Entire floor and all walls will be stainless steel.

- Install interior ladders from Main Level to interior floor.

- Install monitoring camera if desired. Put in opposite diagonal corner of the flood chute.

Phase 20:
Build Maintenance Storage Room

Main Purposes of the Maintenance Storage Room

The primary purpose of the Storage Room is actually not the room itself, but the roof on top of the room. It is on this roof, which is also called the Surface Extension, where the Archimedes Screw will sit. The structure must therefore be able to support the weight of the Archimedes Screw and the water within.

The other important purpose is as an Emergency Container for flood waters. Should the water be greater than the flood container, this room may be opened to contain the additional flood waters.

All other uses are beneficial, but secondary. This is important to remember. The interior of the Storage Room may therefore be reduced in size to maintain the strength of the structure.

The Multiple Purposes of Maintenance Storage Room

The Maintenance Storage Room has numerous purposes, both interior and exterior. The interior is a room where large pieces of equipment and replacement items can be stored. This interior can use used as emergency room for flood management (if the water volume is greater than the Flood Container can handle).

The top of the Maintenance Storage Room is also known as the Surface Extension. This too has multiple purposes. Primarily it can be the base on which the Archimedes Screw will sit. It will also be the location of the Pulley System which brings people and equipment to absolute bottom.

For illustrations see Appendix #3, #6, and #7.

Specific Pictures

The Storage Room has many important purposes, and relationships to other objects. Specific Pictures include: A.3.1, A.3.3, A.3.4, A.3.9, A.3.10, A.6.7, A.7.5, and A.11.10.

These pictures illustrate not only the design of the Storage Room, but the relationship to the Flood Control Room, the Pulley System; and the very important Surface Extension.

General Location and Dimensions of the Storage Room

The Maintenance Storage Room is the other large object at the absolute bottom. It may be larger than the Flood Container.

It is important to note that the Maintenance Storage Room will actually sit on Absolute Bottom, and its top will be equal to the Main Level. This is slightly different from the Flood Container, which sits on a base. Their tops will match in height, but the floor of the Storage Room will be deeper.

Access Doors

There is a door between the two structures. This is a primary method for people and equipment to enter the Flood Container. This door can also be opened remotely in case some water needs to be let out into the adjacent structure.

Note also a ramp is required in the deeper Storage Room in order to reach the door of the Flood Container.

A. Locations and Dimensions of Maintenance Storage Room

- Floor of Storage Room will be at Absolute Bottom.
- All walls built from Absolute Bottom to the Main Level
- Back wall will be adjacent to earth wall up to Main Level.
- Right wall will be adjacent to Left wall of Flood Container.
- Left Wall and Front Wall will be placed accordingly.
- The Top will be same height and area as the Main Level

B. Construction of Maintenance Storage Room

- Build the Floor of Maintenance Storage Room, on Absolute Ground
- Build Back wall and Left wall, up to Main Level
- Build the Right wall, with matching doorway to Flood Container
- The Front Wall and Top will be constructed later

C. Access Doors and Ramps in Maintenance Storage Room

- Build Access Ramp inside Room up to the doorway
  - Gradual ramp for workers and equipment.
  - From floor of room to Door leading to Flood Container.

- Install the door between the Room and Container
  - Sturdy door between Storage Room and Flood Container.
  - Opens inward to the Storage Room.
  - Two doors may be needed, one in each structure

D. Lighting System and Interior

- A modest lighting system can be installed in the Room.
- Powered completely by storage batteries
  - Battery Inside the room or on Top of the room.
- Interior shelving, hooks, and closets as desired.

E. Emergency Opening

- The door(s) between Storage Room and Flood Container can be wired for automatic opening.

- This will be used when extra storage is needed for flood waters.
  - Design of Flood Container and Filter-Pump should be enough in most cases.
  - Yet this provides additional emergency storage of the flood waters when needed.

- Wiring can be done at this time.

- Powered by same power line from surface, using diesel generator, and power line to Main Level.

F. <u>Final Front Wall and Outside Door</u>

- Construct the Front Wall of the Storage Room.

- Create large doorway, to allow any size equipment to enter the Storage Room. Workers and equipment enter through this door.

- Install sturdy door, in case room used as flood container.

- Maintenance Storage Room is now completed.

Phase 21: Install Pulley Systems
Panels, Doors, and Ladders

Now that the Storage Room and Flood Container are completed, it is time to install the Pulley Systems. These Pulley systems will carry men and equipment from the Main Level to Absolute Ground. A pulley system is used because it is simple, and requires only man-power to operate.

In addition, this is a good time to install any remaining doors, panel openings, ladders, and related minor structures.

For illustrations see Appendix pictures: A.3.9 and A.3.10.

A. Install Pulley Systems
- Two Pulley Systems outside the Storage Room.
- Left Side and Right Side, Outside Front Wall.
- Pulleys fixed on Top of Room at Main Level.

B. Install Other Side Panels and Doors
- Other panels and doors can be installed at this time.
- Note other panels will be installed later.

Phase 22: Build Filtration System

The Filtration System can now be built. All equipment for constructing the Filtration System can arrive down the earthen ramp. Yet in the future, for replacing filter layers, the elevator and pulley system will be used.

The Filtration Unit will essentially be a box of 5 layers, with at least three layers of filters. Top and bottom openings allow for the flow in and out of the system. Side panels at each layer will allow easy access to each filter for replacement.

The filter itself, when built, is relatively simple. However, to create this simple filter, we need to build it using the proper order of steps.

For illustrations, see Appendix #3, #5 and #11. Specific pictures include: A.3.4, A.3.10, A.5.1, A.5.2, and A.11.11.

A. Build the Floor and Back Wall

- Build the Floor for the Filter Unit.
    - In front of the Maintenance Storage Room.
    - Right side adjacent to Base under Flood Container
    - Floor as large as needed for Filter Unit

- Build the Back Wall
    - From Absolute Bottom to just above filter hole in Flood Container.

B. Build the Right Wall, adjacent to Flood Container

- Build the Right Wall of Filter Unit, from floor to filter hole opening in the Flood Storage Container.

C. Construct Filter Unit layer by layer, with Left Wall and Panels

- Construct the Filter layer by layer, from bottom upward.

- Filter layers, left wall, and panels will be built as we proceed.

- Front side will remain open while constructing filter system.

- Layers will be, from top to bottom:
    - Water layer 1; Filter layer 1
    - Water layer 2; Filter layer 2
    - Water layer 3; Filter layer 3
    - Water layer 4; Passage to Pump System

- These layers will be built in Reverse order
  - Passage to Pump System
  - Filter layer 3
  - Filter layer 2
  - Filter layer 1

D. <u>Detailed Process of Filter Layer Construction</u> (A.5.1)

- The following is recommended process for Filter Layer Construction.

- Built a short height of Left Wall, to where Filter will sit.

- Install a rectangle support ledge, on all interior sides of the filter unit, which will support the filter.
  - Each filter will rest on the support ledge
  - Ledge thickness and width must support weight of filter, and al water from above.
  - Final ledge on front side will be installed later.

- Create the Filter Box
  - The box will be made of plastic or metal.
  - Mesh tops and bottoms allow water to flow through.
  - Filtering Material will be loaded into the Filter Box.

- Lay Filter Paper on the bottom of each Filter Box
  - "Filter Paper" is a generic term for any one of thin layers of material which are mostly solid, yet very porous, used often for gradual filtration.

  - This Filter Paper will support all of the actual filtration material which will be placed within the box.

- Insert the Filter Box onto the Supporting Ledge.

- Build the Left Wall Piece and Side Panel Door

  - Build the next piece of Left wall, from above this Filter box, to where next filter box will be.

  - Create a side panel while building this piece, to access filter.

- When replacing any one filter:
  - Open the Side Panel.
  - Remove the Filter Box
  - Replace Filter Material
  - Re-insert Filter Box

F. Build the Passage to Primary Pump

- Build the Passageway from bottom of filter to Primary Pump.
- Install a UV Disinfection System on the ceiling of the Passage.
- Panel door on top of passage to replace UV system.
- Power for UV system can be from any of wiring in End Point.

G. Complete the Final Wall of Filter System

- At this time, the final wall (front wall) of filter unit can be created.
- Install all ledges for supporting all filter boxes on this wall.
- Lay the finished wall by raising upright, then secure to other walls.

H. Seal the Filter Unit with the Top

- Place the final pipe from Flood Container to Filter Unit.
  - The pipe will ensure focused water flow to the filter.
  - The pipe will also seal the hole edges, such that water will not leak through around the pipe.
- Build the top of the Filter Unit, to seal it.

I. Test the Filter System

- Test the Filter System for structural integrity.
- Pour water into the Flood Container, watch it flow through the Filter Unit and through Passage Pipe. (Water will be deposited on the Absolute Bottom, in front of the earthen ramp)
- Look for structural integrity, repair any leaks.

Phase 23: Install Primary Pump

The Primary Pump will deliver the purified water from the level of Absolute Bottom up to the Main Level. There are many pumps designs available which will provide this function easily. However there are a few tips and practical points to consider.

For illustrations, see Appendix #3, #5 and #11. Specific pictures include: A.3.3, A.3.4, A.3.10, A.3.8, A.5.2, A.5.3, A.11.11, and A.11.12.

A. Install the Primary Pump (A.3.3, A.3.4, A.3.10)

- Purified water will enter the Primary Pump through the Passageway. The pump will then move the water upward to the Main Level. Purified water will then be deposited into the Archimedes Screw.

- The Primary Pump will be placed in front of the Maintenance Storage Room.

- The Pump will reach from Absolute Bottom to a few feet above the Main Level.

- The Interior volume and pumping power should be able to move large quantities of purified water in a short time.

- Install side doors for men to enter the pump to perform repairs.

B. Install the Motor (A.3.8)

- The Motor for the Primary Pump will be placed at Absolute Bottom, next to the Primary Pump.

- The Power for the Pump will come from the Hydropower System

- A Dedicated Transformer and Dedicated Power Line will carry the power from the Generator to the Pump Motor.

C. Connection to Passage from Filter (A.5.2)

- Make the final pipe connection from the Passageway (from the filter unit) into the Primary Pump.

D. <u>Transferring Water to Archimedes Screws</u> (A.5.3)

- Purified water will eventually be delivered to the second set of pumps: the Archimedes Screws.

- Transfer of water from Primary Pump to Archimedes Screw should be efficient, with minimal loss of water.

- There are two main options: direct pipe, and spout.

- Using a Direct Pipe, the water from the Primary Pump is delivered directly into the Archimedes Screw.

- Alternately, using a Spout, the water is raised above the Archimedes Screw, then poured in from above. Pouring should be from small distance to minimize loss of water.

Phase 24: Build Archimedes Screws

Archimedes Screws as Pumps to the Surface

The second set of pumps are the Archimedes Screws. These will carry the filtered water from the Main Level to the surface, and into the Long-Term Storage Container.

Archimedes Screws are preferred because of their power and simplicity. They can move large volumes of water with minimal electrical power requirements. The rate, while modest, will be fast enough for our purposes.

The Archimedes Screws will be installed at a gradual slope. While at the Main Level, they will be supported by a concrete base. Then while in the earth, the earth itself will hold the Archimedes Screws in place.

Note that the boxes containing the Archimedes Screws will be large, approximately 8 x 8 feet. This size will be sufficient for large diameter screws, plus room for workers to enter for maintenance.

For illustrations, see Appendix #3 and #5. Specific pictures include: A.3.1, A.3.3, A.3.8, A.5.3, A.5.4, A.5.5, A.5.6, A.7.1

Archimedes Screws Built on the Maintenance Room Roof

Note also that the Archimedes Screws will be built on top of the Maintenance Storage Room, rather than the Main Level. It is in fact the main purpose of the Storage Room to be a support for the Archimedes Screws. The storage underneath is a secondary benefit.

It is very important therefore, that the structure of the Storage Room, and the roof in particular, be extremely strong. The Storage Room and the roof itself must hold the weight of 1) the supporting concrete for the Archimedes Screws, 2) the Archimedes Screws themselves, and 3) all of the water being carried upward.

Do not be surprised if the design calculations result in roof and walls which are several feet thick. Support for the Screw System is the primary reason for the Storage Room. Any remaining space permitted inside, however minimal, is a bonus.

Power and Gears for Archimedes Screws

The gearbox for the Archimedes Screw will be at the bottom side of each Archimedes Screw. The power for the Archimedes Screws will come from the Hydropower System. When the Hydropower System is operating, the power will lead directly to the gearbox.

Building the Entire Support System and Archimedes Screws in Open Pit

The Archimedes Screws will require a sturdy support system on the way up. This support will be a solid concrete ramp. The Archimedes Screws will then be built on top of this support ramp.

Eventually the Archimedes Screw will appear to be in three stages: a) from Main Level to Dirt Level; b) through the Dirt Level; and c) above the Surface up and into the Long-Term Storage.

Yet we have the benefit now of constructing the entire thing in an open pit, then sealing later. At this moment, we have an open pit, which allows us to build the entire support system and the entire length of Archimedes Screws at one time.

Later, we will complete the End Point Ceiling, with a hole for the Archimedes Screws. Then we can fill the entire things with dirt. Therefore, we can construct the entire system now, quite easily. Then later perform the seal and ground cover. Dirt will then be filled in around the End Point Structure, filling back up to the Original Surface Level.

A. Build Support for Archimedes Screws

- The Archimedes Screws will be supported by a solid concrete slope.

- Angle of slope can be gentle and long, or steep and short.

- Width must be enough for Archimedes Screws plus a few feet.
    - Extra width will increase stability by spreading weight.
    - Extra width also allows workers to build screw system.

- The concrete slope will be one long structure, yet can be thought about (and perhaps constructed) in 3 sections:
    - From Main Level to Earth Section
    - Through the Earth Section to the Surface
    - From Surface to top of Long-Term Container

- At this moment, we have a large open pit, which allows us to create the Archimedes Screw Supports for the entire length and height.

- We can therefore build the entire concrete slope support at this time, then lay the Archimedes Screws on top.

B. Build Archimedes Screw Box (A.5.3)

- Build the Archimedes Screw Box on top of the concrete support.

- Outer box is concrete, several inches thick each side.

- Inner layer is stainless steel.

- Approximately 8 x 8 ft. minimum.

- Install floor and side walls; leave top open.

- Install access doors and panels on one side of box.

- Multiple Archimedes Screw boxes may be constructed in series.

C. Install Archimedes Screw Blades and Gears (Appendix 3 & 5)

- Install the actual Archimedes Screw Blades and Axel inside the box.

- Install Gearbox and Motor adjacent to Archimedes Screw.

- Power line will come from Hydropower System
  - Dedicated Transformer and Power Line
  - For both Primary Pump and Archimedes Screws

D. Connect Piping from Primary Pump to Archimedes Screws

- Connect the pipe from the Primary Pump to the Archimedes Screw

- Pipes will either connect to Archimedes Screw from the side at the bottom, or from above.

- Piping must be secure, such that the pressure of the water will not displace the piping and cause interior flooding.

- Test the Archimedes Screws using Battery Power and small water volume, before sealing.

E. <u>Seal the Archimedes Screws</u>

- Build the lower end of Archimedes Screws with concrete wall.

- Leave the far end open for water to pour into surface container.

- Seal the box with a sealed top.
  - Concrete or thick metal as top of Archimedes Screw Box
  - First 2 feet length of roof is metal door, for necessary access.

- Double check all piping connections and seals.

- Double check all wiring.

Notice that Next 3 Phases are Double Checks

The next 3 Phases are actually Double Checks. At this time, all of the End Point Structures should have been created. We are now double checking some of the details, before we create the final seal of the entire thing.

Phase 25: Finalize All Pumping Connections and Seals

This Phase serves as a Double Check. The entire system has numerous pipes, tunnels, pumps, and so on which carry immense volumes of water. We need to double check everything.

At this moment, go through the entire system in the End Point and finalize the entire water transmission system. Install any remaining pipes or pumps as needed. Double check the seals on every connection.

Do this for everything within the End Point. The time spent now to double check...and seal properly...will save much heartache in the future. Should one piece burst apart, the entire End Point may become flooded.

Phase 26: Finalize All Electrical Wires for Internal Use

As with the previous Phase, this Phase is also a Double Check. There are many Power Lines installed throughout the End Point. We have planned appropriately to install each power line as we built the End Point. Therefore most of the power lines should be installed. Yet we must make sure of this.

At this time, we will go through the entire Power System of the End Point. We will make sure every location which requires power has its appropriate power line, and that this power line is connected to its proper power source. The following are a few Reminders.

A. General Usage of Each Power Source

- Hydropower is used to operate those items needed during the Flood Event. This includes: Pumps of Filtered Water; and Main Power Line for the Public. (A.3.6 and A.3.8)

- Solar Power is used for most of Day Use Operations, such as Maintenance near the Entrance Points. (A.9.2)

- Batteries with Solar Power is used for those areas close to surface, yet need to be used during flood events. This includes lights, doors, and antennas at the Entrance Points. (A.9.2)

- Batteries with Hydropower is used for those items which can be powered by Hydro, yet may be used during the flood event or during maintenance. This includes much of the lighting and doors in the Maintenance Tunnel. (A.3.8)

- Diesel power is used where we absolutely need a certain amount of power, and we can get this from the surface level (primarily above the End Point). This includes: Elevator, Ventilation, and Main Lighting. This also includes large equipment operation during maintenance. (A.8.7)

- Stand-Alone Batteries are used where there are single locations, difficult to reach with an extensive power line (simpler if we don't), and with minimal power requirements. These include: Lighting for Storage Room; Lighting for Absolute Bottom; and some Surveillance Cameras.

B. Make sure that each of the following has power lines
- Motor for Primary Pump
- Motors for Archimedes Screws
- UV Disinfection
- Main Power Line to the Public

- Lights above Main Level
- Lights in Storage Room
- Extension power line from surface
- Elevator

- Lights in Maintenance Tunnel
- Maintenance Doors with Electronic Locks or Remote Operation
- Surveillance Cameras
- Any Mechanical Ventilation Systems

C. Hydropower will Provide Power for the Following Directly
- Motor for Primary Pump
- Motors for Archimedes Screws
- Main Power Line to the Public
- UV Disinfection

D. Hydropower with Batteries will Provide Power for Following
- Lights in Maintenance Tunnel
- Maintenance Doors with Electronic Locks or Remote Operation

E. Surface Diesel Power will Provide Power for the Following

- Lights above Main Level
- Extension power line from surface
- Elevator

F. Local Solar Battery Power Will Provide Power for the Following

- Lights in Maintenance Tunnel
- Maintenance Doors with Electronic Locks or Remote Operation
- Radio Wave Antennas
- Some Surveillance Cameras

G. <u>Higher Voltage Transformers Provided For</u>
- Motor for Primary Pump
- Motors for Archimedes Screws

H. <u>Supplemental Power and Back-Ups</u>
- Supplemental power can be provided from alternate power sources and alternate power lines.

- Power lines from solar, hydro, and diesel can be linked in some places, turned on when needed.

- These can be switched on at any time.

- Examples include: Elevator, Lights, Doors, Ventilation.

Phase 27: Construct and Lay
the Main Power Cable on Tunnel Roof

In the Order of Steps above, we have given the opportunity for laying the Main Power Line earlier. However, engineers may choose to wait to install this power line until all items and all power lines are put together within the End Point. If this Power Cable was not installed earlier, then it should be installed now. (See drawings A.2.11 and A.11.6).

As a reminder, the Main Power Cable will provide emergency electrical power to the general public. This power comes from the Hydropower System.

The amount of power from the Hydropower is divided into priorities. It is first sent to power the Primary Pump, and the Archimedes Screws. Most of the remaining power will go to the Main Power Cable. A smaller amount will be sent to storage batteries, then to the Emergency Escape Tunnel.

Local engineers will have determined the size of the turbines used, and the amount of power which can be generated. This will result in dedicated generators and dedicated transformers for each of the purposes. First priority will be to the pumps; second priority is to the Main Power Line the Public.

A. Reminders on Laying Power Cable on Maintenance Tunnel Roof

- Main Power Line is sent from back of Generator Room to the Maintenance Tunnel, and onto the Roof.

- The Cable is laid on top of the Maintenance Tunnel Roof, following the slope of the Tunnel Roof, toward the city center.

- Cable is then brought to surface and connects to existing power lines where desired.

Phase 28:
Build the Long-Term Storage for Clean Water

The Long-Term Storage Containers

On the surface at the End Point will be a series of Long-Term Storage Containers. These containers will store the purified water, above ground, as long as may be needed. Purified water will then be accessible to the public, from each of these containers around the city. For illustrations, see Appendix #5 and #8.

Structure and Material of the Containers

The design of these containers is simple, yet a few details are important. The structure should be sturdy, likely made of concrete. The interior should be sanitary, therefore lined with either a layer of hard plastic or stainless steel.

We can either use one large container, or a series of containers connected in series. The total volume must be the same as for the short term flood container below ground. (See Appendix #8)

Filling the Containers (A.5.4, A.8.2)

The first Long-Term Container is filled from the Archimedes Screw. The Archimedes Screw lifts the water, and pours into the container.

The second container is filled via connection pipes. These pipes are installed approximately 2/3 from the bottom of each container. When the water in the first container fills to that level, the water will flow through those pipes, and into the second container.

A similar set of pipes can be placed between container #2 and #3. Ideally, this is placed just a few inches lower than the previous pipes on the other wall. Thus, when the container #2 fills to that level, the water will then flow into container #3. This provides a cascading waterfall effect which can done for up to 5 Long-Term Containers.

Meanwhile, the Archimedes Screw continuously drops water into container #1, which keeps the cascading process on-going, until all the water has been deposited. Note that the final amount of water in each container will be approximately the same based on this design.

Additional UV Disinfection System

An additional UV Disinfection system can be installed in each of the Long-Term Containers. Each UV system will be installed as a row of UV lights on the celling of the container. The UV will help disinfect the water while the water remains in the container, adding an extra degree of purification.

The power for the system will be provided by solar cells, which sit on top of each container. Batteries will also be stored on the roof, in boxes which are protected from the weather. Solar cells load electricity into the batteries, which then cause the UV system to operate.

Additional Filter Options inside the Container (A.8.4, A.8.5)

Additional filters can be placed inside the long-term storage containers. We can use the cascading flow of water from one container to the next to not only move the water among containers, but to use multiple filters along the way.

Water drops into a filter unit, becomes extra purified, and then is sent out a pipe in the bottom. From here, the water will gradually fill up the long-term container. The filter unit will essentially become submerged within the purified water of the container.

This process can continue for each long-term container. The pipes between two storage containers will not only allow water to flow to the next container, but will also pipe the water into the next filter. The water will be filtered – again – becoming more purified. The water is sent out the bottom pipe, and gradually fills up the storage container. And so on.

Note that to make this work, the Filter Unit must be very sturdy, as it will eventually be submerged in water. We do not want the unit to break. Furthermore, we must have pipes and funnels which will carry the water directly into the filter.

Accessing the Purified Water

The purified water will be accessed from two pipe extensions on each of the long-term containers.

The first pipe extension will allow trucks to fill with purified water, then deliver to the neighborhoods where the water is needed most. The second pipe extension will permit connection to the existing municipal water supply. Both pipes will remain closed until needed.

A. Placement and Sizing of Long-Term Storage Container(s)

- Location of the First Storage Container is co-designed with the length and slope angle of Archimedes Screw. (A.8.7)
  - Design both before building the Archimedes Screw
  - Length and Slope of Screw will put water at desired location

- The First Long-Term Storage Containers should be placed near the pit, but not directly over the pit.

- No container will ever be placed on the area above the Underground End Point. (A.8.7)

- The First Long-Term Storage Container should be a minimum of 50 feet from the edge of the pit.

- The series of Long-Term Containers will likely be built in a direction approximately 90 degrees from the Flood Chute direction.

- The Total Volume of all Long-Term Storage Containers together must be equal to (or greater than) the volume of the underground flood storage container.

B. Build the First Long-Term Storage Container

- Begin building the First Long-Term Storage Container.

- One corner directly under the top of the Archimedes Screw.

- The Wall under the Screw will be shorter than all other walls.

  - A short rectangular hole will soon exist between this wall and the roof of the container.

  - The top of the Archimedes Screw will essentially rest on this wall, and partially into this hole.

  - The water will then pour directly into the first container through this slit.

  - Note also that the roof will protect the container from rain, which is not part of our purified water system.

- Floor and Walls are made of Concrete.

- Interior Stainless Steel or other sanitary material.

- On the wall opposite of Archimedes Screw install a row of connection pipes through wall to the next Storage Container.

- Connection pipes are placed approximately 2/3 from bottom.

C. Build Each Adjacent Storage Container

- Construct each sequential Storage Container.

- Each one adjacent to the next, in a line.

- Connection pipes between each set of containers.

- Height of next set of pipes a few inches lower than the previous.

D. Install Exit Pipes as we Build the Walls

- Install exit pipes on each storage container as you build.

- Best to place these exit pipes on wall 90° from the wall with connector pipes.

- Extend the pipes 3-6 inches beyond container.

- Place approximately 2-3 feet from ground.

- Seal each tightly; removed only when needed.

E. Add the Roofs, UV System, and Solar Power

- Lay the concrete roof for each container.

- Install the UV Disinfection system in the ceilings of Containers.

- Build the Solar Array on the Roofs.

- Set the storage batteries in weather-proof boxes, and connect all the wiring.

F. Roof System for Rainwater Collection and Independent Filter

- Install the gutter system to drain water from the roof.

- Place simple filter and container on side to collect purified rain water.

- These side containers will provide additional purified water, which can be siphoned into containers and taken to the public.

G. Option for Filters inside the Long-Term Containers (A.8.4)

- An additional option for filters is to place them *inside* the Long-Term Containers.

- The design for series of containers above results in a cascading effect of water into each container. This process can also be used in consecutive filtration units within the containers.

- Use Gravity Filtration Units in each Long-Term Storage Container.

- Make sure that each filtration unit is solid. The exterior of the filter unit must not break, or be penetrated by the water within the storage container.
    - The Filter Unit will eventually be submerged in water, as the storage container fills up. Therefore the filter unit must never crack or break from the force of the water.

- Connect wide funnel and pipe, directly from Archimedes Screw to the first gravity filter. Make sure all of the original water falls into the funnel and pipe, to be led directly to the first filtration unit.

- Install similar pipe connectors from each filter in the other containers to the transition pipes in the walls.

- The process will begin as water dropping from the Archimedes Screw into the funnel, through the pipe, and into the first Filter Unit.

- After filtration, the water exits the filter unit at the bottom of the filter (this is also the bottom of the storage container). This water will gradually fill up the container from the bottom.

- When water reaches the level of the transition pipes in the far wall, the water will flow into the next storage container. This water will flow directly through the connector pipe to the filter unit below.

- Again the filtered water exits the filter at the bottom of the storage container, and gradually fills up the container…until reaching the level of the pipes on the far wall.

- The process then repeats for each long-term storage container.

H. Additional Filter outside First Long-Term Storage Container

- An additional filter option is to place a gravity filter outside the first storage container, prior to entering the storage container.

- Place this filter directly under the Archimedes Screw. The Archimedes Screw is already raised above ground, and can therefore pour water into the gravity filter (rather than the storage container).

- After the water passes through this filter, then the water can be pumped into the first Long-Term Storage Container.

- This will be an additional pump, which can be powered by the solar power and batteries on top of the first storage container.

Phase 29:
Construct Elevator Room on Surface

The Elevator Room can now be constructed on the surface. This is an item that can wait until near completion of the entire construction project, and that time is now.

The elevator itself (shaft, elevator, power) has likely been completed already. Constructing these parts is easiest to do while still doing the construction at the main level.

Yet this elevator can remain in rough appearance for the remaining phases. Being in rough appearance will simplify any working operations while construction continues. However, at this time, we can construct the final Elevator Room. (A.3.8, A.8.7)

A. Build the Elevator, if not done already

- If Elevator has not been built, do so now.

- Elevator should be placed beyond the area of the pit.

- Doors underground should face either Main Level or Tunnel.

- Power will come from diesel generator on surface.

- This is a freight elevator, with wide dimensions, for large equipment and replacement parts.

- Public will use as escape route up, but not as entrance down.

B. Construct the Elevator Room on Surface

- The Elevator Room will be a simple structure.

- Concrete box structure, which surrounds elevator.

- Metal doors, which are locked to public.

- Diesel generator to the side supplies the power.

- Manual switch inside room connects elevator power.

- Similar manual switch in Tunnel also connects power.

- Rain gutter on roof channels water to pipe on opposite side of diesel generator (to protect generator). Can lead to rain barrel if desired.

Phase 30:
Build Supporting Walls and Pillars on Storage Room

Structural Support for Surface Above

Remember that eventually we will enclose the entire Underground End Point in concrete, and cover with soil. Therefore we will need structural support which will hold the concrete roof, and the soil above.

We will build this into parts. First we will build the support that will sit on top of the Storage Room. Then we will build the support that will be built from Absolute Bottom.

For illustrations, see Appendix A.7.3 and A.7.4.

Left Wall and Pillars on top of Storage Room / Extension Shelf

We need support for the area above the Storage Room. However, we must also remember that much of this area will remain open. Therefore it will not be a box, but rather a left wall and a set of pillars.

The left wall can be a solid wall. Build this wall adjacent to the dirt of the pit.

On the right, just a few feet before the Flood Storage Container, we will have a set of pillars. These pillars will have similar support as a wall, yet be more open. Pillars may also be placed elsewhere on the Main Level if needed for support.

The rest of the area will be open, without any walls or pillars.

A. Build the Supporting Walls and Pillars above Storage Room

- Build the Left Wall, of concrete, next to dirt of pit wall.
- Build a set of Support Pillars along the right side.
    - A few feet from the Flood Container.

Phase 31:
Build End Point Structure Walls
From Absolute Bottom to Surface Level

The End Point Structural Boundaries: General Reminders

Eventually we will seal the entire Underground End Point in concrete. We refer to this as the "Structural Boundary" of the End Point. The concrete will protect the Underground End Point from all potential damage from the earth.

Furthermore, we will bury the entire End Point under several feet of soil. Therefore the roofs that we place over the Structural Boundary must be permanently stable. This means that the roof slabs, and the support for those roof slabs, must be strong enough to hold the earth above essentially forever. For illustrations, see Appendix #7 and #11.

Specific drawings include: A.7.1 through A.7.5 (All of Appendix #7 is related to the Structural Boundaries); plus A.11.12 and A.11.13.

Regions Requiring Physical Support

There are two regions which should have completed support at this time. The first is the Main Level, which should have the support walls, as well as the actual roof. The second is the support for the area above the Storage Room. This was completed in the previous phase, and the roof will be added later.

In this phase we will build the largest support structures: the walls from Absolute Ground to the Surface Level. We will build solid concrete walls from the Absolute Ground to the Surface Level. These walls will support the roofs, which will then support the earth above.

Designs for Structural Boundary of the End Point and Wall Placement

Note that there are two general designs for the Structural Boundaries. The first design is to create a large box, which surrounds the entire End Point. The second design is to have the walls of the Main Level be the final width, and continue with that in our final walls.

a) If we build a large box around the entire End Point, then we must first plan for this by digging a pit which has extra dimensions on all sides. Note that our Main Level will be sitting on a large area of dirt. Also, the Flood Container will be sitting on a base of dirt.

The visual effect will be similar to cutting a cake, a layered cake, very smoothly. We trim this "cake" on all sides. Then, we build a surrounding set of walls. Each wall will be the height from absolute bottom, to the surface level. Each wall be long enough to enclose the entire End Point.

b) The second design will have each section be side walls in themselves. Thus, the walls and pillars of the Main Level will be the first section of Left and Right Walls. Then the walls and pillars on the Storage Room will be second section of Left and Right Walls.

The final left and right walls will simply be an extension of those walls. The width of the final structural boundary having been determined by the previous sections, we simply continue with that width. Therefore, at Absolute Ground, our left wall will extend from the Storage Room left wall. The right wall at absolute ground, will similarly extend from the right wall of the Flood Storage Container

Remove All Trucks and Equipment

This is the time to remove all trucks and large equipment from the area. Bring them back up to the surface level using the earthen ramp, before beginning this phase.

Strategy for Access when Building

Some strategy should be considered before beginning this phase of construction. Eventually we will have enclosed the End Point, with very tall concrete walls. Consider how the men and equipment will come in and out of the work area.

The primary method is of course the same method workers will use when the project is complete. This is to use the freight elevator to get to the main level. Then use the pulley system to get down to the absolute bottom level.

Another method is plan for this phase early, by making the pit much wider and longer than the final dimensions. This will allow worker to go down the ramp, and walk around the outside of the walls. The men can therefore build the walls from the outside of the walls, rather than from the inside.

These methods can of course be supplemented by cranes on the surface bring men and materials to the appropriate depth.

A. <u>Build Left and Right Walls from Absolute Bottom to Surface Level</u>

- Build Left and Right Walls from Bottom to Surface.
- Reinforced Concrete Walls.
- Height from Absolute Bottom to the Surface Level.
- Length from Dirt Walls to in Front of Primary Pump.
- Note that Floor remains Dirt for any leaking water to soak into.

B. <u>Length of Left-Right Walls Depends on Design of Structure</u>

- There are two designs for the Structural Boundaries, and therefore for the positions of the Left and Right walls:

    a) The Large Box Room, where a set of long and high walls surrounds the entire End Point.

    b) The walls of each section are the final width of the End Point, and we extend the walls from there.

- Choose the Design of the Structural Boundary, then place the final Left and Right Walls accordingly.

- Make sure the final left and right walls are thick enough to support the concrete roof and the earth above.

C. <u>Build Front Wall from Absolute Bottom to Surface Level</u>

- The Front wall should the last wall constructed.

- Place approximately 25 feet in front of Primary Pump.
    - The horizontal line from this distance is final position for the Front Wall.

- Build from Absolute Bottom to the Surface Level.

D. <u>Seal End Point with Final Roofs</u>
- See next Phase

Phase 32:
Lay End Point Roofs to Seal End Point

Filling the Ramp to be Equal to Roof Level

It is now time to seal the Underground End Point. We begin by filling the dirt of the ramp, raising it to the level of the roofs which we will install. Now we still have a ramp, but it is much closer to the surface. Just as important, the ramp will lead directly to the top of the walls. For illustrations, see Appendix A.11.12.

At this point, we can easily construct the roof sections on top of the structural boundary walls. (See Appendix #7 and #11).

Sections of Roofs to Construct

There are two sections of roofs to construct. The first one is above the Storage Room. The second is above the Absolute Ground.

The roof above the Storage Room should itself be installed in sections, as it will be built around the existing Archimedes Screws and ventilation pipes. The final result will be that the Archimedes Screws and ventilation systems poke through holes in the roof (then extend through the topsoil to the surface).

The roof above the Absolute Ground (far above absolute ground) can be constructed in a single piece. However, it is recommended that the roof here be constructed in 5 to 10 equal sized segments. These will be thinner sections, and therefore inherently more stable. These can be laid parallel, similar to wood planks on a floor.

The roof sections can be installed in any order.

Time for Concrete to Fully Harden

It is extremely important to give the concrete full time to harden. The walls must be fully cured before roofs are installed. Further, the roofs must be fully cured before being buried under the soil. Under no circumstances will these processes by rushed. Too many engineering projects have fallen apart based solely on rushing this step. Do not let that happen here.

A. <u>Fill Dirt to Raise Ramp to Roof Level</u>
  - Re-fill the earthen ramp with the soil previously removed.
  - Raise the end point of ramp equal to height of roof placement.
  - Trucks and equipment will drive directly to top of Front Wall.

B. <u>Build Roof above the Absolute Ground Area</u>
  - Build the roof above the Absolute Ground Area.
  - Drive trucks directly to the top of Front Wall.
  - Lay the roof sections in segments or as one large piece.
  - Install additional Support Pillars where needed.

C. <u>Build Roof around Archimedes Screw</u>
  - Build roof sections around the Archimedes Screw.
  - Also build around ventilation pipes.
  - Ensure entire roof system above End Point is complete.

D. <u>Add Gutter Extensions for Rainwater to Flow Away from Structure</u>
  - Add gutters around the entire rim of the entire roof of End Point.
  - All gutters will lead to the Front Wall.
  - Final gutter pipe extends outside the Front wall, down that wall, to the earth below.
  - The purpose of this gutter system is to prevent rain damage from the earth above.
    - Rain above the End Point will naturally sink into soil above the End Point, and into the earth.
    - Eventually rainwater will land on roof of End Point.
    - To prevent damage over time, this rainwater is channeled off the roof, to outside the wall structure, And into the ground beyond.

Phase 33:
Replenish the Dirt and Grass

When the entire Structural Boundary of the End Point is complete, including the entire roof, then we can re-fill the dirt and bury the End Point Systems.

We have already started this process by filling up the earthen ramp, to reach the top level of the Front Wall. Now we continue, and fill all areas of the construction site. When we are done with this phase, there will be nothing visible. All that remains is a dirt area above the End Point.

A. Fill Outer Regions with Rocks, Dirt, and Soil (A.11.13)

- The Outer Regions, beyond the Walls of the End Point Structure can now be filled.

- Re-fill with material previously dug to make the pit.

- Begin with rocks at the bottom; then clay soils; finish with nutrient rich soils at the top.

B. Complete Fill of Dirt to Original Surface Level (A.11.13)

- We can now re-fill the entire construction site with soil.

- The ramp will of course be eliminated.

- Use the material previously dug to make the pit as fill.

- Gradually bury the entire End Point under several feet of soil.

- Use the softer soils as much as possible for the main fill.

- Use nutrient rich top soil as the top 3 feet (minimum).

- Fill all construction site back to Original Surface Level.

C. Reclamation of End Point on Surface (A.11.13, A.8.7)

- The only visible reference to the construction site will be a layer of dirt over the area above the End Point.

- To complete the reclamation, plant grass over the area, and install a few solar arrays.

D. <u>Using Any left-Over Dirt and Rocks</u> (A.8.7)

- There may be some rocks and soil left-over from the project.

- Use this material for artificial creeks, berms, and small hills.

Phase 34:
Finalize Ventilation Structures on Surface

There will be ventilation systems allowing air into the Underground End Point. On the surface, at this moment, they stick out of the ground as pipes. We need to take these pipes sticking out, and finish the ventilation systems above ground. This stage is the best time to finish the ventilation systems.

Passive Ventilation with Protection from Rain

Most of these ventilation systems will be passive, as air naturally flowing from surface to the underground. However, we need protective coverings. These must be protected from rain coming in, and yet allow the air to circulate. There are two ways to do this.

The first method is use a curved pipe on the surface. Air can flow up and into the curved pipe, yet the rain will simply fall off the pipe to grass beneath.

The second method is to use a tent cover with mesh walls. The tent cover is simple metal, and is sits on top of the pipe. Water will flow over and down. The tent will be above the pipe, and therefore allow air to circulate in, but water to drop off. This can be also improved by enclosing in a mesh container, with a tent shaped top. Again, the water flows off the tent, yet the mesh allows air to circulate.

Active Ventilation Systems

Active ventilation systems can also be installed. These will be powered by diesel generators or wind power.

The most important point is the weight of the system, and therefore the placement. The power source and the motor will have significant weight, and therefore must not be placed directly on the End Point. Instead, the power and motor should sit far from the End Point (50-100 feet).

The Ventilation Motor should sit on the opposite side of the Archimedes Screws. This is important for weight balance near the End Point. It is also important to prevent contamination of ventilation motor chemicals with the purified water.

Beyond those points, local engineers can make their own choices. There are many types of active ventilation systems. The engineers will determine best design, size, and so forth.

Phase 35:
Beautification of End Point Meadow

At this point, we have completed most of the tasks for constructing the End Point. In fact, we have completed most of the projects for the entire Flood Control System. For the Meadow area at the End Point, all that remains is Beautification, and a few final details.

For illustrations see Appendix #A.8.7.

Beautification Engineering Details

The End Point will be primarily a Meadow. The entire surface area will be a meadow of simple grass. Only a few structures (such as the elevator and long-term storage) will be sitting on this meadow.

Note that there are a few important engineering details regarding beautification. The most important point is to *not* add weight or interfere with the underground. Specifically this means:

- No structures on top of the End Point (except solar cells)
- No trees near the end point (the root structure will cause problems)
- No playground equipment or goal posts.

Solar and Wind Power

Solar Power and Wind Power can be placed on the Meadow. The meadow is a large, flat area. Therefore it is a perfect place to install large sets of Solar Arrays. These solar arrays can be placed almost anywhere in the meadow.

Similarly, we can also install wind turbines. It is best to place these wind turbines in the far corners of the meadow. Make sure they are far away from the actual End Point structures. Only 1 or 2 turbines will be enough.

These solar cells and wind turbines can provide regular electrical power for the nearby neighborhoods. However, do not rely on them for the Flood Control System.

Artificial Creeks

Artificial creeks can be installed along the meadow. The purpose of the artificial creek is to channel the heavy rains which fall there, on the meadow, to a specific location.

Trees and Shrubbery

Trees and shrubbery may be planted, but very minimally. This area should remain primarily an empty meadow.

If planted, these should be at the edges on the meadow, along the border. These should be far away from all aspects of the End Point structures.

Security

Security is important for our End Point. We do not want anyone to willfully damage the Structures of the Flood Control System. We also do not want anyone to be able to enter the underground and cause serious damage within the Underground End Point.

Therefore, the Elevator Room (which is the primary access point to the underground) will have a sturdy metal door, with a strong lock. This can only be opened by authorized personnel.

Furthermore, the entire End Point Meadow should be surrounded by a fence. A simple chain fence should be sufficient, gate entrances. These gates will be locked with heavy chain. Managers will decide when these gates will be allowed open.

Security cameras may also be installed at various locations around the meadow.

Public Use of the Meadow

Depending on the location, the public may be allowed to use the Meadow. However, there are some important conditions.

First of course the meadow must be surrounded by a fence, with a locked gate. The meadow will remain locked at night, and only opened during the daytime.

The meadow will also remain a simple meadow. There will be no playground, sports facilities, or picnic tables. It will be a meadow and nothing more.

It is also important to consider the neighborhood. If the public of the area is generally peaceful and good-natured, then the meadow can become open to the public. If, however, the neighborhood has people who are vandals and criminal in nature, then the meadow should remain closed to the public.

There will be only three people who have the keys to this meadow: local park managers, power technicians, and maintenance crew managers. It is their discretion when the meadow will be open or closed to the public.

A. Plant Meadow Grass

- Cover the construction site with nutrient rich top soil.
- Plant meadow grass throughout the entire meadow.

B. Install Solar Cells and Wind Turbines

- Install Large Solar Arrays anywhere in the meadow region.
- Install Modest Wind Turbines on the far edges of meadow.
- Connect the power to local neighborhood power systems.

C. Create Creeks, Hills, and Berms

- Using left-over earth from construction, build artificial creeks and other topography.
- Topography should be designed for water flow and soil preservation during heavy rains.

D. Build Access Roads

- There will be two access roads for trucks and equipment.

- The First Access Road will run parallel to where the Underground End Point is located.
  - Gravel Road (stable, yet allows water to soak in).
  - Wide 2-Door Gate on the right, to enter Meadow.
  - Short dirt or gravel road to the End Point Structures.
  - No admittance on this road except for crews.

- The Second Access Road will run parallel to the meadow along any of the other sides.
  - This road provides quicker access to wind turbines.
  - May also be a small neighborhood or county road.
  - Simple locked gate for power crews to enter.

- The other two sides of fenced meadow can be bordered by:
  - Greenbelt, Creek, or Small Park
  - Farms and Farm Roads
  - Small neighborhood streets.

E. Install Security Fences, Locks, and Cameras

- Build a security fence around the entire perimeter of the End Point Meadow.

- Place the fence poles deep into the ground, with concrete support.
  - This is important to ensure the fence will remain in place, never knocked down, during the heavy rains.

- Install each gate, with secure chain locks.

- Public access gate will only be from greenbelt or neighborhood side.

- Build a sturdy metal door to the Elevator Room, with a chain lock.

- Install security cameras around the meadow as desired

The End Point Meadow is Now Complete

Phase 36:
Build or Finalize Entrance Point Structures to the Maintenance Tunnels

At this point, most of the Flood Control System has now been completed. All Flood Chutes, all Maintenance Tunnels, and the entire End Point Structures, have all been completed. We can now leave the End Point Meadow permanently.

What remains is to finalize the Entrance Point Structures. These are the buildings and staircases which allow access to various locations in the Maintenance Tunnel. These were likely built and finished earlier, but if not, then now is the time to complete them.

For details on the steps, see various Phases above. For illustrations, see Appendix #9. Here are a few brief reminders.

A. Build the Entrance Point Shafts and Staircases

- Dig the shafts from surface to tunnel.

- Install the staircase.

- Make sure the staircase is easy to walk without slipping.

- Install batteries and wiring.

B. Build the Surface Structures of Entrance Points

- Build a raised mound for water to drain off.

- Surround the shaft with concrete base.

- Enclose in concrete structure.

- Build angled roof for rain flow and solar cells.

- Install solar cells and radio antennas.

- Install gutters and rain barrel

- Install locked door and surveillance camera.

    - Will be unlocked each storm.

- Beautify the surrounding areas.

Phase 37:
Test the System

Importance of Testing the System

Now that we have completed the construction of the Flood Control System, it is time to test the system.

This system, as designed, should work perfectly. Furthermore, we have tested and made repairs along the way. However, there may be sections which are not constructed perfectly. It is better to learn of these now, and make adjustments, rather than finding out during the next major flood.

The primary objective for these tests is to ensure that every section of the system can handle the volume of water effectively. The flow, the pressure, will put strain on the tunnels and pipes. The continuous pressure will thrust against every seal and joint. We must therefore test the durability of each structure and each seal to ensure that the water pressure will not cause any damage.

Primary and Secondary Testing of the System

The primary importance of testing the system is to ensure the Drainage Rings, Flood Chutes, Connecting Pipes, and Seals are structurally perfect. From street level to flood container, water must be able to flow perfectly without creating any cracks or broken seals.

Yet there is a secondary set of equipment to inspect. These are for the Hydropower System and the Pumping System. Watch the turbine operate and the generator create power. Observe the flow of water through the filter, primary pump, and Archimedes Screw. Follow the cascading flow of water in the long-term containers.

When all the equipment, structures, and seals have passed the test, then the entire Flood Control System can be officially opened.

Method for Testing the System

The basic method for the system is to artificially inject large volumes of water directly into the Entrance Grates. Furthermore, water will be injected at multiple points simultaneously. This will simulate actual flood conditions. After the test, workers will manually inspect the system, and make repairs as necessary.

More specifically, it is best to test the system spoke by spoke. It is also important to test all Drainage Rings along that spoke simultaneously. This will provide maximum volume for testing at the End Pont.

Coordinate Water Injections

Ideally, the way to do this is to coordinate the water injections. The timing should be done sequentially. Every time water in the Flood Chute passes a Drainage Ring, water should also be flowing from the Drainage Rings into that Flood Chute. This will approximate actual flood conditions.

Thus, every time the water in the Flood Chute passes a Drainage Ring, more water is added to the water in that Flood Chute. This is how it will be in actual flood conditions.

By the time the initial water has passed all Drainage Rings, all the other water injections have joined in the Flood Chute. This total volume of water, with the force of the water on the walls, is what the Flood Chute must be able to handle. Similarly, the Turbine, Flood Container, Filter Unit, and Pumps, must be able to handle this volume (and pressure) easily.

Engineers can easily calculate when to inject each volume of water into Drainage Rings. This is based on distance traveled, and speed of water. The calculations do not need to be perfect, but close enough to make the test work properly.

Re-Using Water

All of the water can be re-used. Remember that the water will eventually be pumped into the long-term storage containers on the surface. Therefore, each of the trucks can reload their containers with the same water, then drive to their location along the next spoke to perform the next test.

This process can be done for the testing of all structures and equipment in each spoke of the system, one spoke at a time.

Therefore, it does not matter that we are using large amounts of water for testing the system, because the water will be reused multiple times.

Specific Tips for Testing on the Next Page

A. Test Planning for Flood Control System

- Do the tests one spoke at a time.

- Inject into all Drainage Rings along one spoke in each test.

- Water will be injected directly into Entrance Grates above each Drainage Ring.

- Select the Entrance Grate geographically closest to Flood Chute, for each Drainage Ring. All injections are done at these Entrances.

- Place each truck with a large volume of water at the designated Entrance Grate.

- Estimate sequence of water injection such that water will merge in the Flood Chute.
    - Water from each Drainage Ring will merge with water already in flood chute.
    - This simulates larger volumes of actual flood conditions.

- Place men within the End Point to monitor flood control operation at the End Point.

- Place men along the Maintenance Tunnel, near the Drainage Rings, to monitor water flow at those locations.

B. Water Volumes and Pressures for the Tests

- Use clean water.
    - No dirt, oil, or chemicals.
    - These may be involved in real floods, but not in testing.

- Inject as much water as is practical at each Entrance Grate.

- Inject water at the highest pressure possible.

- Greater volumes of water, and at greater pressure, will be a more accurate test of flood conditions.

- We will obviously not be creating a full flood, yet simulating as close as possible to actual conditions, for short period, will help us see any places which need strengthening.

C. Performing the Test

- Place all water trucks and men into position.
- Timing Manager radios each team when to start next injection.
- Crew in Tunnels and End Point watches for any problems.
- Test completed when all of water is in Long-Term Container.
- Crew will report any issues.
- Final team at surface containers will report when test is complete.

D. Evaluation During the Test

- During the test, crew will evaluate the following:
- Rate of flow of water past their area.
- Major structural problems.
- Turbine and Generator effectiveness.
- Rate of Filtration.
- Effectiveness of Primary Pump and Archimedes Screws.

E. Evaluation After the Test

- After the test, crew will manually evaluate the following:
- Each Drainage Ring.
- Entire length of Flood Chutes, all walls and floor.
- All equipment at the End Point.
- Looking for cracks and damaged seals.

F. Repairs and Adjustments

- Make any repairs and improvements as needed.
- Ensure Tight Seals and Structural Stability
- Modify for Effective Power Generation and Distribution

G. Preparing for Next Test

- Refill the trucks with water from the Long-Term Containers.

- Drive trucks to each of their next designated points.

- Position crews along next tunnel and End Point stations.

- Use the experiences from previous tests to make modifications in the next test.

Phase 38: Create Maintenance Schedule

This system will last for at least 300 years, and likely much longer, if the structures are maintained properly. Therefore we must set up an on-going schedule of inspection and repairs for this system.

The Flood Chutes and Drainage Rings are the most important structures to inspect and repair. Inspection of all sections of chutes and pipes must be inspected annually; and always after a major flood.

- Create an Inspection and Maintenance Schedule for the entire Flood Control System.

- Inspection and Maintenance can be done section by section, spoke by spoke, over time.

- Inspection should be on-going repeated sequence of each section.

- Primarily repair seals and cracks.

- Also inspect power systems.

- All sections of Drainage Rings and Flood Chutes must be inspected annually.

- All End Point equipment inspected minimum of twice each year.

- Filters replaced based on experience with materials.

- Inspection and Maintenance of the entire system must also be done after each major flood.

Phase 39: Project Completed
Virtual Tours and Grand Opening Ceremony

Congratulations! The entire Advanced Flood Control System has been completed. You can trust that this system will protect many lives, and prevent miles of property damage, for several generations.

If maintained properly, this system will function for at least 300 years. You citizens will be very happy knowing they are protected by this system.

Virtual Tours and Scale Models

I recommend creating Virtual Tours and Scale Models of the Flood Control System. The people will want to know how it works. They will want to know what it can do for them. Therefore, we can create virtual tours and scale models for the people to see.

(A) Virtual tours are professional videos which can be placed on official websites. Professional videographers can be hired, and escorted, into the various parts of the systems. These videos will be narrated by engineers involved in the project.

The result will be a series of short videos (30 minutes maximum) which show the various parts of the Flood Control System. Particular emphasis should be given to two concepts: 1) what this part of the system will do for the people, and 2) how this part of the system should be maintained. (The maintenance emphasis will be important to keep adequate funding for proper maintenance).

These videos can be placed on official websites, such as county and city government sites, as well as public sites such as YouTube.

(B) Scale models can also be built to help the public understand the operations of the system. Small scale replicas of the parts can be placed on a replica of the territory, and placed under protective glass. These can be placed in city libraries, utility headquarters, and science museums.

(C) Several 3-D graphic programs can also be created. These are a combination of the virtual tours and the scale models. Using 3-D graphic imagery, the people can take a very virtual tour. It will be as if walking through the system. These programs can be made available for viewing at science museums and city libraries.

Limiting Access to the Actual System

It is also important to remember that the public should not be allowed in these facilities. We do not want the system to become vulnerable to vandals or terrorists. Therefore, virtual tours and scale models are the best way to show off the system, without compromising security.

The Grand Opening Celebration

Creating this Advanced Flood Control System for the people is a major accomplishment. Everyone involved in the project should be extremely proud. In fact, the entire region should be celebrating!

There should be a full day celebration at the Control Center. In nearby streets, parking lots, and parks, the community should celebrate the opening of this Flood Control System.

The Control Room should have a large ribbon. The chief engineers, construction manager, and city leaders who championed the system should be up there...say a few words...and cut the ribbon. The system is official open.

Although the public cannot be allowed into the system itself, they can be allowed into the areas of the control room, and nearby buildings.

A media room should be set up, to show the overview of the system, in 20 minute film. Scale models can be placed in the building or in the lot outside. Any photos taken during the project can be enlarged and displayed.

Outside should be a fair type atmosphere. Use your imagination. Just make it a happy, family style celebration. Make it something the entire community can enjoy, and remember for years. This project benefits the entire region, and all the people of the region should feel part of the celebration!

Selling Commemorative Items

Of course you will want to make and sell commemorative items. Shirts, coins, pamphlets, and toy models. Anything which the public may want to buy, to commemorate the Opening Day.

The public will want to buy these commemorative items. It will make them feel more connected. Looking upon the item, the person will remember the joy of the day. They will also be proud of this new flood control system in their region.

In return, the managers will recoup a small amount of cost of the construction. However, above all, this is a Celebration for the Community! The public must feel joy and pride from the day spent at the Opening Fair!

Additional Resources and Information

Additional information on the Flood Control Designs can be found in several places.

1. Abridged Version of Advanced Flood Control System
- Available as e-book and print book, from Amazon.

2. Full Version of Advanced Flood Control System
- Available as e-book and print book, from Amazon.

3. Amazon Author Page
https://www.amazon.com/author/markfennellvisionary

4. Main Website for all Writings, Links, and Pictures
https://markfennellvisionary.com

5. Main Blog
http://markfennell.blogspot.com/

6. Videos on You Tube: "All Things Energy"
http://www.youtube.com/channel/UCk5ckPqF4oD0JoJMSBi2Zcg

7. Facebook
https://www.facebook.com/mark.fennell.758

About the Designer

My name is Mark Fennell. As a scientist and inventor, I have modeled myself after such greats as Leonardo Da Vinci and Nikola Tesla.

Regarding the design of this Flood Control System, know that I have studied civil engineering for years. This includes water management, mining operations, tunnel construction, structural stability, park design, and all aspects of electrical power.

I have examined many engineering designs from history (ancient, medieval, and modern); particularly water management and concrete structures.

I also have practical experience with the Red Cross Disaster Service. This includes assisting with Disaster Preparedness Plans and participating in Disaster Response Training. I have organized databases and libraries for disaster-related materials; during which time I read numerous disaster histories and response plans.

Also note that I have personally experienced flooding in Houston, and Dallas; throughout much of Ohio; and several areas of Indiana. I am also familiar with the rivers and drainage systems in Germany and Eastern Europe. I know the geography, the topography, and the existing inadequate drainage systems in each of those areas.

Be confident that this design will work, and will prevent any future flooding in metropolitan areas. When installed and maintained properly, this system will be effective.

A few Accomplishments worth Noting:
- Authored and published over 60 books
- Created over 30 inventions
- Discovered solutions which eluded greatest scientists

Most of my Scientific Discoveries and Inventions are related to:
- Electrical Power (All Components)
- Energy Science and Gravity
- Motion and Friction
- Water Flow and Filtration
- Unified Energy and Quantum Science

You can learn more about my discoveries and books at:
http://markfennellvisionary.com

M.F.

Contact Designer

Most information can be found in the various books and on-line resources. However, if additional information is needed, the designer may be contacted on Facebook or email. Facebook and Messenger is the preferred method.

FB: https://www.facebook.com/mark.fennell.758

email: markpoet@aol.com.

www.ingramcontent.com/pod-product-compliance
Lightning Source LLC
Chambersburg PA
CBHW081728220526
45468CB00008B/2018